Holt McDougal Mathematics

Course 2
Chapter 9 Resource Book

HOLT McDOUGAL
a division of Houghton Mifflin Harcourt

Printed in the United States of America

ISBN 13: 978-0-55-400746-5
ISBN 10: 0-55-400746-0

2 3 4 5 170 12 11 10 09

Contents

Description of Contents

Family Involvement Pages

The Chapter Resource Book includes a set of family involvement pages for each section. The family involvement pages consist of the following items:

- **Family Letter**, a two-page letter describing the math that the student will study in each section. A list of vocabulary words from the lessons is included, and some worked-out examples are provided.
- **At-Home Practice**, a one-page worksheet with problems drawn from the content described in the Family Letter. Answers are provided on the page so that the student and his or her family can check this work at home.
- **Family Fun**, a one-page activity sheet that the student and his or her family can work on together.

Practice A, B, and C

There are three practice worksheets for every lesson. All of these reinforce the content of the lesson. Practice B is shown in the Teacher's Edition and is appropriate for the on-level student. It is also available as a workbook (Homework & Practice Workbook)

Practice A is easier than Practice B but still practices the content of the lesson. Practice C is more challenging than Practice B.

Review for Mastery

The Review for Mastery worksheet (one per lesson) provides an alternate way to teach or review the main concepts of the lesson. This worksheet is one or two pages long and is shown in the Teacher's Edition.

Challenge

The Challenge worksheet (one per lesson) enhances critical thinking skills and extends the lesson. This worksheet is shown in the Teacher's Edition.

Problem Solving

The Problem Solving worksheet (one per lesson) provides practice in problem solving and opportunities for real-world applications and for interdisciplinary connections. There are both multiple choice and short response problems. This worksheet is shown in the Teacher's Edition.

Reading Strategies

The Reading Strategies worksheet (one per lesson) provides tools to help the student master math vocabulary or symbols.

Puzzles, Twisters & Teasers

The Puzzles, Twisters & Teasers worksheet (one per lesson) provides fun practice while reinforcing the content of the lesson.

CHAPTER 9

Family Letter

9A Perimeter, Circumference, and Area

Dear Family,

The student will be learning about accuracy and precision in measurement. In the real world, no measurement is exact and all measurements are approximations. **Accuracy** is the closeness of any given measurement or value to the actual measurement or value. **Precision** is the level of detail an instrument can measure. The smaller an instrument can measure, the more precise its measurements will be. For example, a millimeter ruler has greater precision than a centimeter ruler.

Choose the more precise measurement in each pair.

A. 1 m or 104 cm
   A centimeter is a smaller unit than a meter, so 104 cm is the more precise measurement.

B. 5 gal or 5.2 gal
   One tenth of a gallon is a smaller unit than a gallon, so 5.2 gal is the more precise measurement.

The student will study the perimeter of figures and circumference of circles. The **perimeter** of a figure is the sum of the side lengths of the figure or the distance around a figure. **Circumference** is the distance around a circle.

The student will use these formulas to find the perimeter of a rectangle and the circumference of a circle.

Perimeter of a Rectangle	Circumference of a Circle
$P = 2\ell + 2w$ ℓ = length w = width	$C = 2\pi r$ or $C = \pi d$ r = radius d = diameter

Pi is approximately equal to 3.14 or $\frac{22}{7}$ and is represented by the symbol, π.

Find the circumference of the circle to the nearest tenth. Use 3.14 for π.

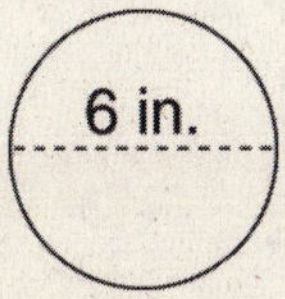

$C = \pi d$	You know the diameter.
$C \approx 3.14 \cdot 6$	Substitute 3.14 for π and 6 for d.
$C \approx 18.84$	Multiply.

The circumference of the circle is about 18.8 inches.

Vocabulary

These are the math words we are learning:

accuracy the closeness of a given measurement or value to the actual measurement or value

area the number of unit squares needed to cover a figure

circumference the distance around a circle

composite figure a figure made up of simple geometric shapes, such as triangles and rectangles

perimeter the distance around a geometric figure

pi the ratio of circumference C to diameter d $\left(\frac{C}{d}\right)$ for every circle; approximately equal to 3.14 or $\frac{22}{7}$

precision the level of detail an instrument can measure

CHAPTER 9

Family Letter

9A Perimeter, Circumference, and Area continued

The student will also be learning how to find the **area** of different types of figures. He or she will use the following formulas to find the area of these figures.

Area Formulas	
Rectangle 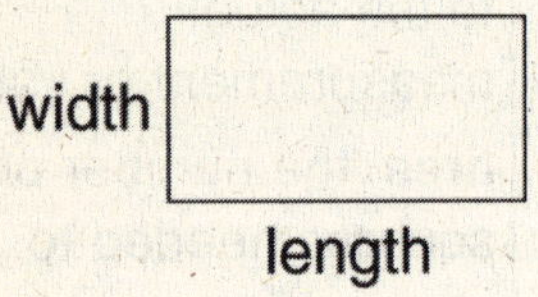	$A = \ell\omega$ Area = length • width
Parallelogram 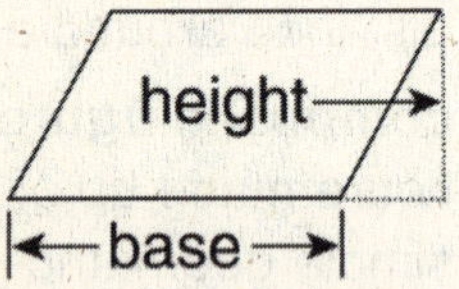	$A = bh$ Area = base • height
Triangle 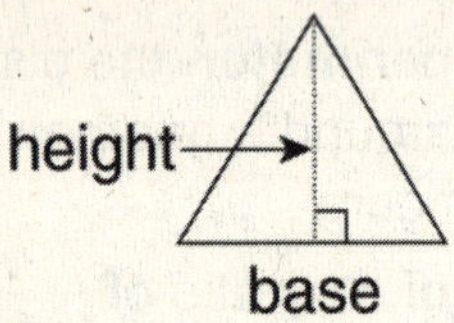	$A = \frac{1}{2}bh$ Area = $\frac{1}{2}$ • base • height
Trapezoid 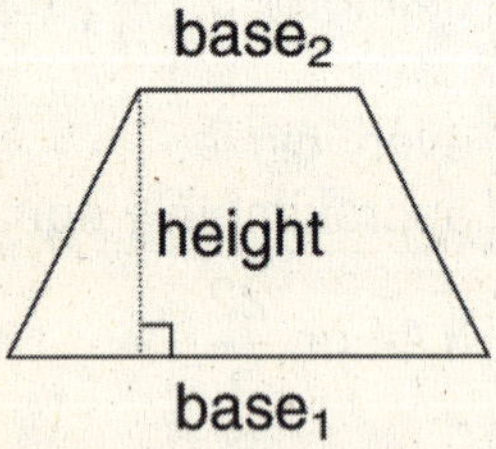	$A = \frac{1}{2}h(b_1 + b_2)$ Area = $\frac{1}{2}$ • height • (base$_1$ + base$_2$)
Circle 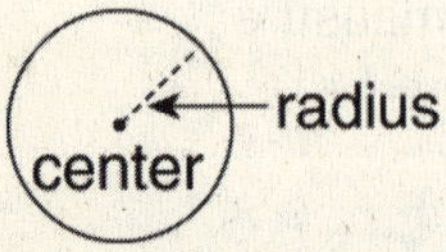	$A = \pi r^2$ Area = π • radius^2

Discuss with the student the difference between finding area and perimeter. Work with the student to calculate the area of different figures in your home.

Sincerely,

Name ______________________ Date ______________ Class ______________

CHAPTER 9

At-Home Practice

9A Perimeter, Circumference, and Area

Find each perimeter.

1. 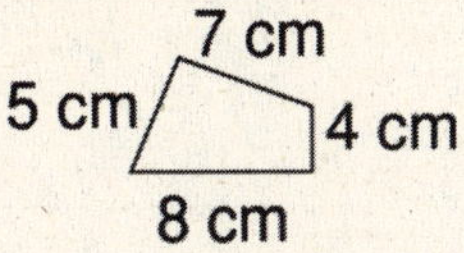

2. 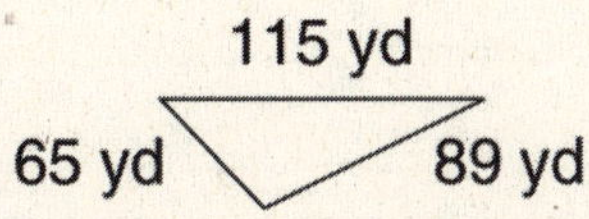

3. 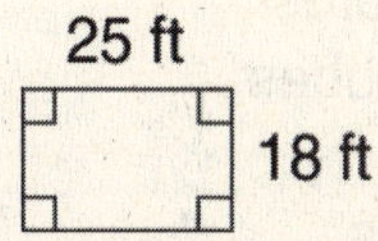

Find the circumference of each circle to the nearest tenth. Use 3.14 for π.

4. 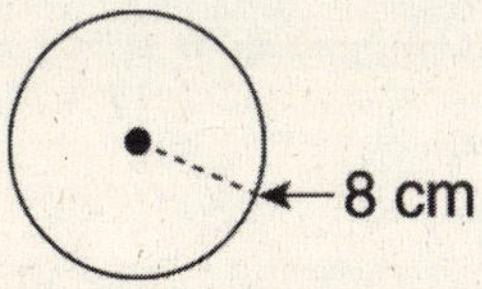

5. 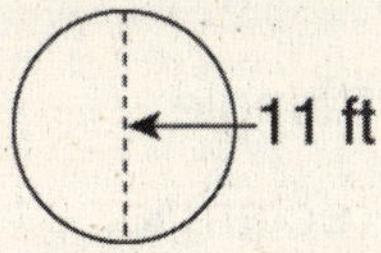

6. 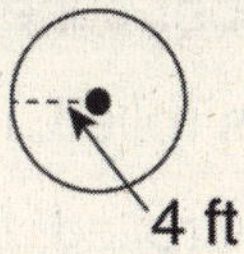

Find the area of each figure.

7. 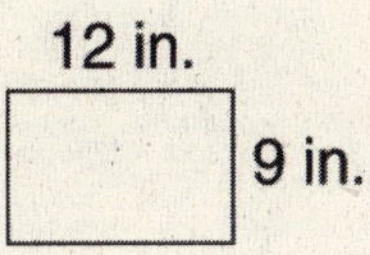

8. 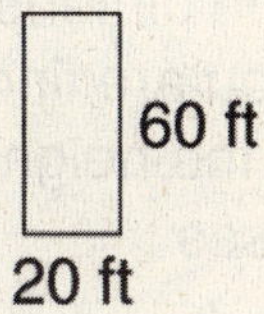

9. 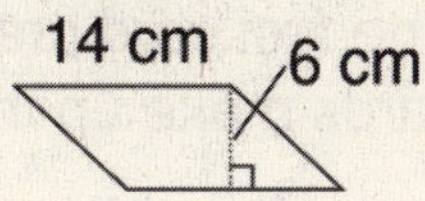

10. 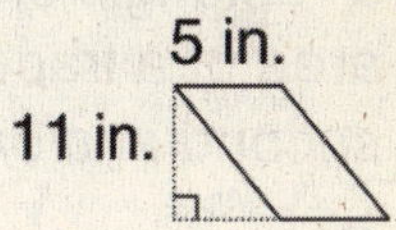

11. 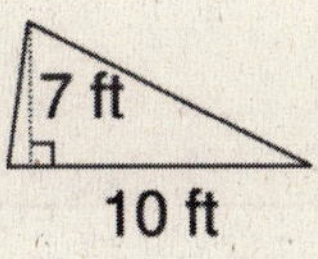

12. 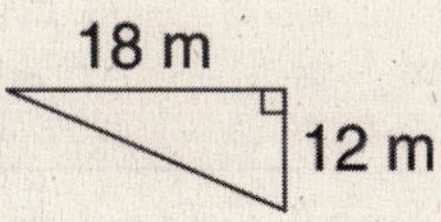

13. 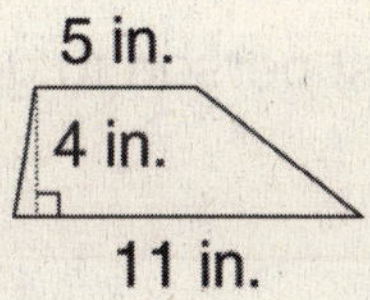

14. 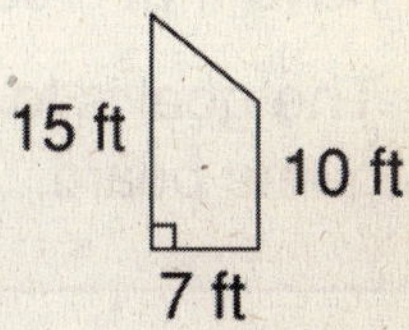

Find the area of each circle. Use 3.14 for π.

15. A circle has a radius of 24 inches. What is the area? Round your answer to the nearest tenth.

____

16. A circular fountain has a radius of 14 feet. What is the area of the fountain? Round your answer to the nearest tenth.

____

Answers: **1.** 24 cm **2.** 269 yd **3.** 86 ft **4.** 50.2 cm **5.** 34.5 ft **6.** 25.1 ft **7.** 108 in^2 **8.** 1,200 ft^2 **9.** 84 cm^2 **10.** 55 in^2 **11.** 35 ft^2 **12.** 108 m^2 **13.** 32 in^2 **14.** 87.5 ft^2 **15.** 1,808.6 in^2 **16.** 615.4 ft^2

Name ______________________________ Date ______________ Class ______________

CHAPTER 9

Family Fun

Area Cross Out

Materials

Deck of cards

Number cubes

Directions

- Write any non-prime whole number from 1 to 100 in each box. These numbers represent the area of any triangle, rectangle, or trapezoid.
- Roll the number cube.
  - If a 1 or 2 is rolled, find the area of a triangle.
  - If a 3 or 4 is rolled, find the area of a rectangle.
  - If a 5 or 6 is rolled, find the area of a trapezoid.
- The deck of cards will provide the values for the dimensions. Each card will hold its own number value. The face cards can be any even number. You will draw 2 cards at a time for the dimensions of a rectangle or a triangle. You will draw 3 cards if you are finding the area of a trapezoid. The first card drawn is always the height. The second and/or third cards make up the base or bases.
- Once you know the figure for which you are calculating the area, draw the correct number of cards and calculate the area. If you have that area on your game board, cross that square out.
- The goal is to be the first person to get a "four in a row" on your game board.

Name ________________ Date ________________ Class ________________

LESSON 9-1

Practice A

Accuracy and Precision

Circle the more precise measurement in each pair.

1. 2 mi 1,200 yd
2. 84.2 oz 84 oz
3. 3 m 200 km
4. 12 fl oz 12 c
5. 11 pt $11\frac{1}{2}$ pt
6. 25 ft 35 yd

Measure to the nearest half, fourth, and eighth inch. Which measurement is the most accurate?

7. 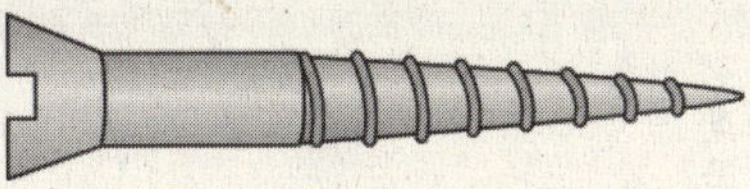

____

8. 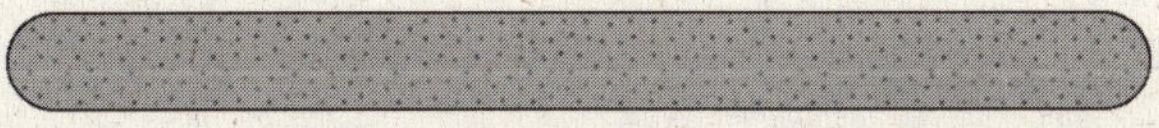

____

Estimate each measurement.

9. 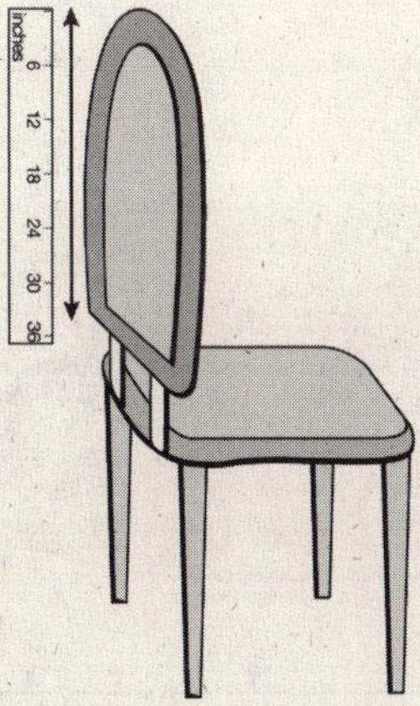

____

10. 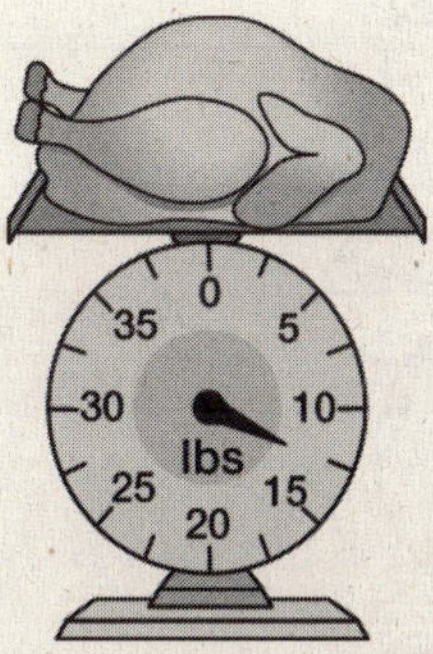

____

11. Anchorage, Alaska, averages 15.9 inches of rain per year. The town of Annette, by contrast, gets 8.6 feet. Which measurement is more precise? Explain.

____

____

12. Typically, Bettles, Alaska, averages 12.5 inches of snowfall in January and 0.11 inches in August. Which measurement is more precise? Explain.

____

____

Name ______________________ Date ______________ Class ______________

LESSON 9-1

Practice B

Accuracy and Precision

Circle the more precise measurement in each pair.

1. 122 km 1,230 m
2. 91.1 kg 91 kg
3. 14 cm 14 mm
4. 5,281 ft 5,281 in.
5. 1 mi 1,000.3 mi
6. 550 g 5 kg

Measure to the nearest half, fourth, and eighth inch. Which measurement is the most accurate?

7. 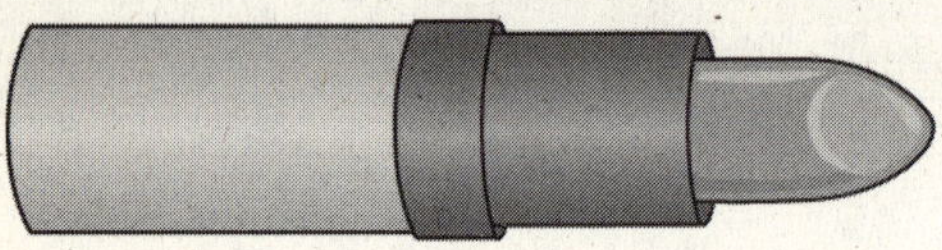

____

8. 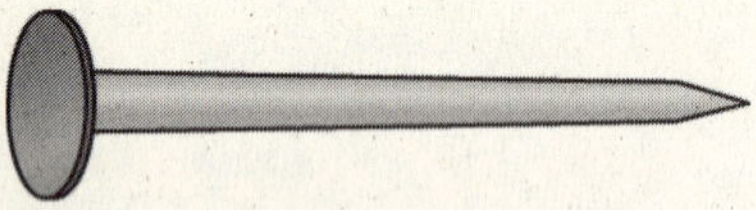

____

Estimate each measurement.

9. 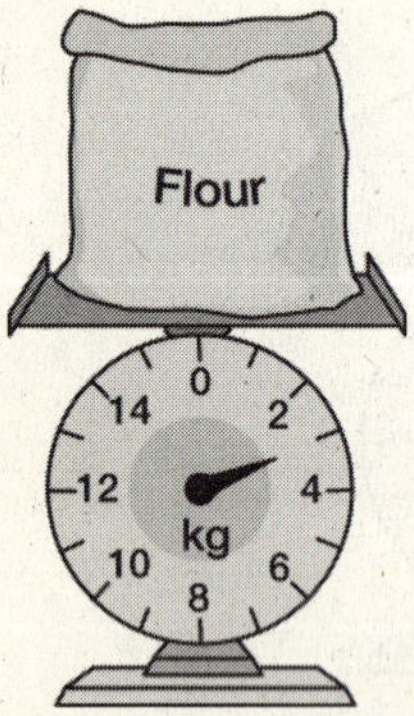

____

10. 

____

11. Mt. Washington, New Hampshire, averages $8\frac{1}{4}$ feet of rain per year. Las Vegas, Nevada, gets 4.13 inches. Which measurement is more precise? Explain.

____

____

12. In a recent August, the town Portage, Indiana had 16.67 inches of rain. The normal amount of rainfall in August is about 4 inches. Which measurement is more precise? Explain.

____

____

Name ________________________ Date ______________ Class ______________

LESSON 9-1

Practice C

Accuracy and Precision

Circle the more precise measurement in each pair.

1. 1,500 m   15,001 km
2. 91 °F   90.4 °F
3. 18.75 mi   1.5 mi
4. 17 oz   2 cups
5. 1.0 qt   1.11 qt
6. 3,451 cm   3,451 mm

Measure to the nearest half, fourth, and eighth inch. Which measurement is the most accurate?

7. 

____

8. 

____

Estimate each measurement.

9. 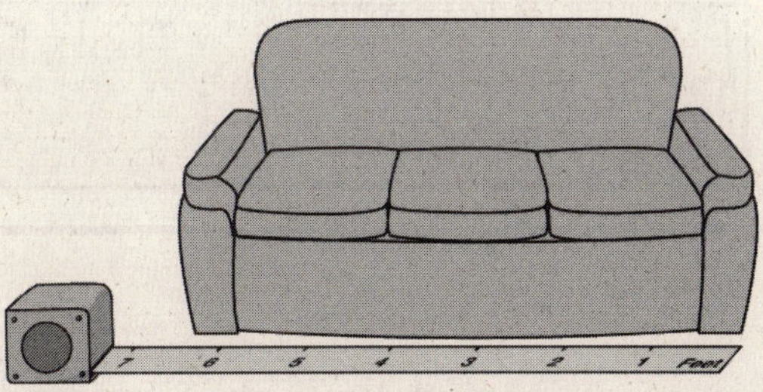

____

10. 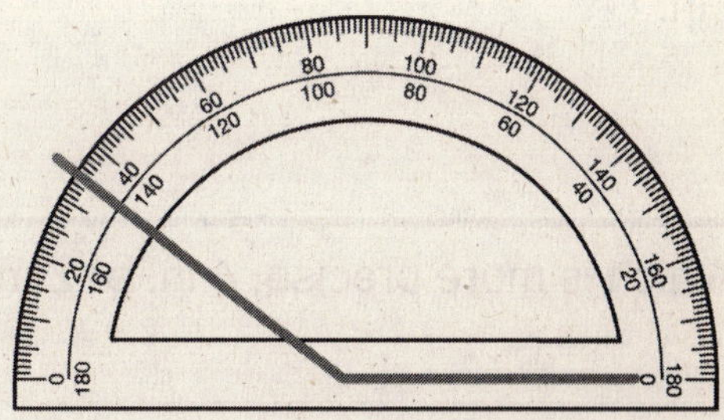

____

11. Valdez, Alaska, averages 27.3 feet of snowfall per year. The town of Yakutat, by contrast, gets 201.4 inches. Which measurement is more precise? Explain.

____

____

12. Mt. Shasta, California, averages 29.9 inches of snowfall in January but only 9.0 inches in November. Which measurement is more precise? Explain.

____

____

Name ______________________ Date ______________ Class ______________

LESSON 9-1

Review for Mastery

Accuracy and Precision

Some measurements are more **precise** than others because the **unit of measure** is smaller. A standard ruler is divided into several units of measure. The ***smaller the unit***, the ***more precise the measure***.

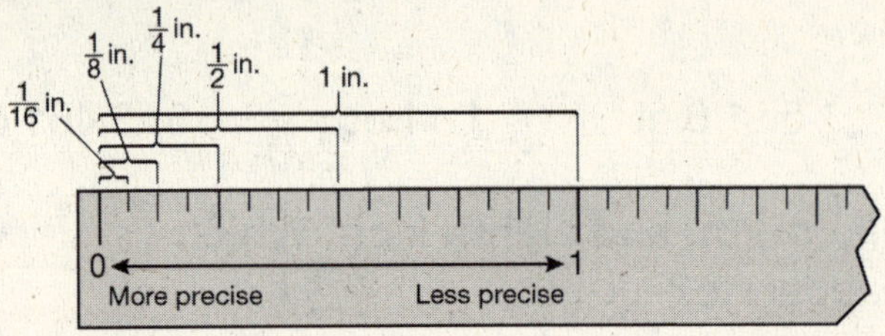

You can extend this idea to other units of measure. In each example, the more precise measurement is circled:

12 ft, (13 in.) — ***inch*** < ***foot*** *more precise*

17 km, (120 m) — ***meter*** < ***kilometer*** *more precise*

14 oz, (14.2 oz) — ***tenth*** < ***one*** *more precise*

Accuracy is how close a measured value comes to the real value:

So, the most accurate measurement is $2\frac{7}{8}$ in.

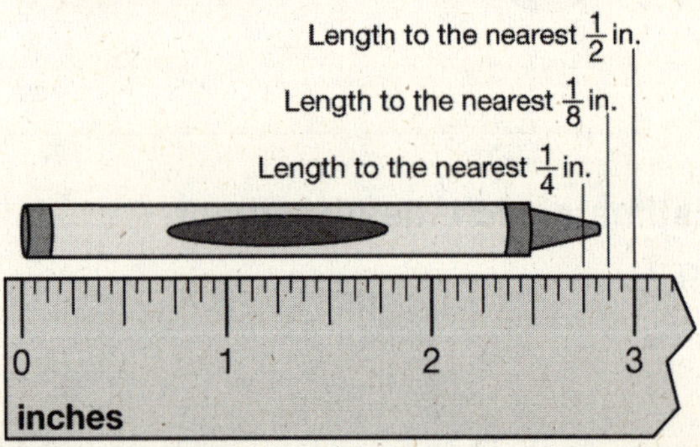

1. Which is more precise, 4 in. or 2 miles?

____

2. Which is more precise, 2 mm or 2 m?

____

3. Which is more precise, 2 mm or 2.1 mm?

____

4. Measure the length of the address label to the nearest half, fourth, and eighth inch. Which measurement is the most accurate?

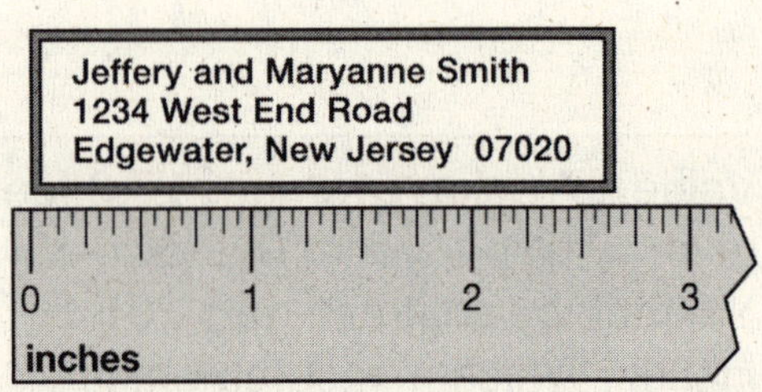

____

Name ______________________ Date ______________ Class ______________

LESSON 9-1

Challenge

Errors in Measurements

No real world measurement is exact. The accuracy depends on the precision of the instrument used. In some cases, the level of error makes little or no difference. For example, an error of 1 meter is not important if you are finding the average distance from Mars to Earth. For an architect designing a house, however, an error of 1 meter could be very serious.

The degree of error of a given measurement is known as **experimental error**. Also called *percent error* or *relative error*, experimental error can be expressed as an equation:

$$\text{experimental error} = 100 \times \frac{(\text{observed value} - \text{accepted value})}{\text{accepted value}}$$

Suppose, for example, you use a high-precision tool to check the volume of a one-cup measure. You observe that the volume is 7.6 fluid ounces instead of the accepted value of 8 fluid ounces. To find the experimental error, use the above equation:

$$\text{experimental error} = 100 \times \frac{(7.6 - 8)}{8}$$

$$= 100 \times \frac{(-0.4)}{8}$$

$$= \frac{-40}{8}$$

$$= -5$$

The result is expressed as a percent: the cup measure is “off” by –5%

Find the experimental error.

1. You examine a ruler using a high-precision tool and discover that 1 in. is actually 0.912 in.

   ____

2. You examine an analog scale using a high-precision tool and discover that 1 lb is actually 15.886 oz.

   ____

3. You examine a 2-liter bottle using a high-precision tool and discover that it actually holds 1,994.7 mL.

   ____

4. You examine a yardstick using a high-precision tool and discover that 1 ft is actually 11.915 in.

   ____

5. You examine a radio telescope that measures light years. You discover that the device is calibrated so that 1 light year = 63,709 astronomical units (au) instead of 63,240 au, the true equivalent.

   ____

Name ______________________ Date ______________ Class ______________

LESSON 9-1

Problem Solving

Accuracy and Precision

Solve the problems.

1. Carol is measuring her room to center her new bed along one wall. Will her measurements be more precise if she uses centimeters or inches? Explain.

____

____

2. Normal rainfall in Hilo, Hawaii is 2.36 feet per year. Yearly, rainfall in Honolulu, Hawaii is 7.77 inches. Which measurement is more precise? Explain.

____

____

3. The Akashi-Kaikyo Bridge in Japan measures 6,529 feet. The Great Belt Bridge in Denmark measures 1,624 m. Which measurement is more precise? Explain.

____

____

4. Santa Barbara, California, gets an average of 1.03 feet of rainfall every year. The city of Santa Maria gets 13.95 inches. Which measurement is more precise? Explain.

____

____

The drawing shows salad dressing ingredients. The top liquid is oil, the bottom vinegar. Circle the correct answer.

5. Estimate the amount of vinegar in the cup.

A. $2\frac{1}{2}$ fl oz C. $3\frac{1}{2}$ fl oz

B. 3 fl oz D. 4 fl oz

6. Estimate the amount of oil in the cup.

A. $\frac{3}{4}$ fl oz C. $3\frac{3}{4}$ fl oz

B. $\frac{2}{3}$ fl oz D. 4 fl oz

7. Estimate the amount of water it would take to fill the cup to the 6-oz level.

A. $1\frac{1}{4}$ fl oz C. 3 fl oz

B. $2\frac{1}{4}$ fl oz D. $6\frac{3}{4}$ fl oz

Name ______________________ Date ______________ Class ______________

LESSON 9-1

Reading Strategies

Compare and Contrast

Precision and **accuracy** both relate to measurement. However, there are differences:

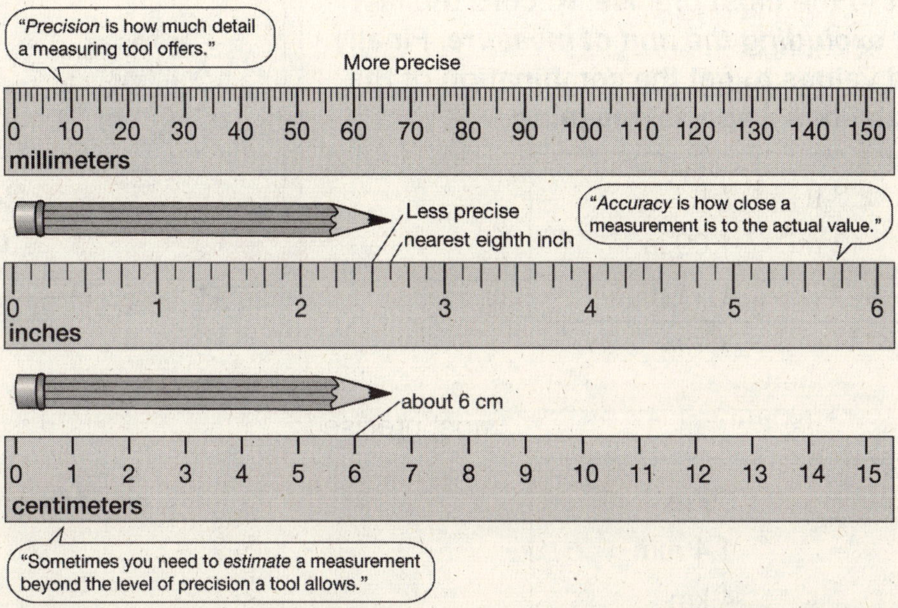

Sometimes you need to **estimate** a measurement beyond the level of precision a tool allows.

Which measurement is more *precise*?

1. measurement in inches; measurement in feet

   ____

2. measurement in yards; measurement in miles

   ____

3. Which is the most accurate measurement of the pencil: $2\frac{1}{2}$ in., $2\frac{3}{4}$ in., or $2\frac{5}{8}$ in.?

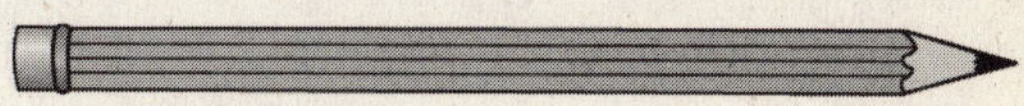

____

Name ______________________ Date ______________ Class ______________

LESSON 9-1

Puzzles, Twisters & Teasers

What's the Combination?

You can figure out the combination by ordering each group of measurements starting with the *least* precise and moving right to the *most* precise. Record the last numerical value, *excluding the unit of measure*. Finally, add the recorded values to get the combination of the safe. *Hint:* The total has a special meaning.

1.

1.36 mi	2.3 ft	7 ft
1.2 in.	43 mi	1.09 in.
17.1 mi		1.11 ft

______________________ most precise :

2.

1 mm	99 m
4.9 k	1.4 mm
28 cm	4 km

______________________ most precise:

3.

3 gal	5 tbsp	0.6 tsp
8.5 fl oz	2.44 qt	53.4 qt
9 pt		1 fl oz

______________________ most precise:

4.

0.2 mg		501.2 kg
6,000.3 g	100 kg	111 mg
0.05 mg		6,000 g

______________________ most precise:

The combination is ___ . ___ ___

Name ______________________ Date ______________ Class ______________

LESSON 9-2

Practice A

Perimeter and Circumference

Find the perimeter of each polygon.

1.

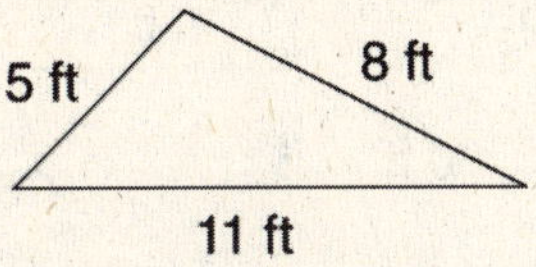

____

2.

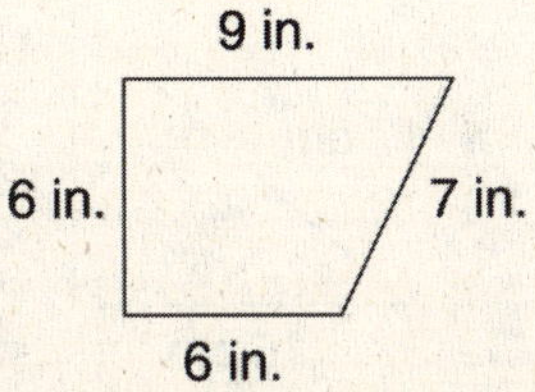

____

3.

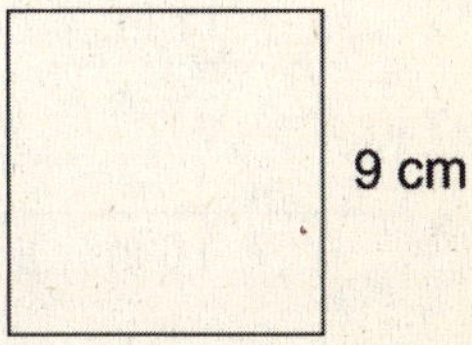

____

Find the perimeter of each rectangle.

4.

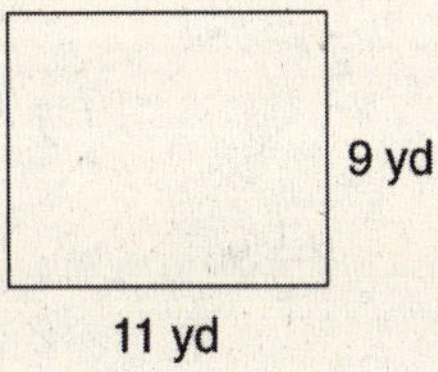

____

5.

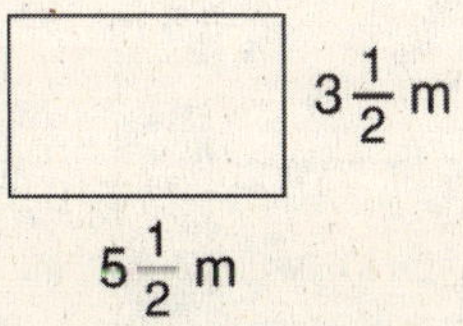

____

6.

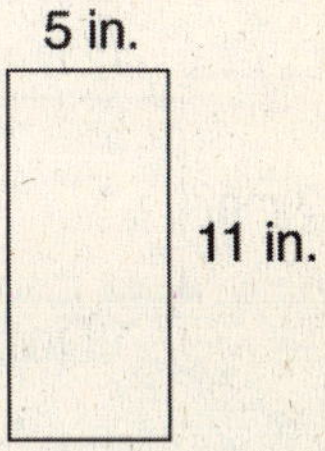

____

Find the circumference of each circle to the nearest tenth. Use 3.14 for π. Choose the letter for the best answer.

7.

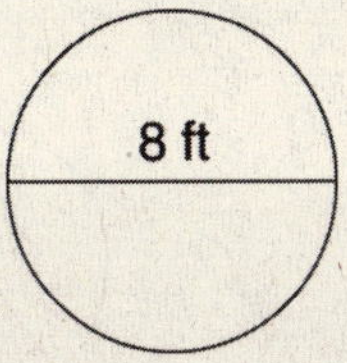

A 12.5 ft C 25.1 ft
B 12.6 ft D 25.2 ft

8.

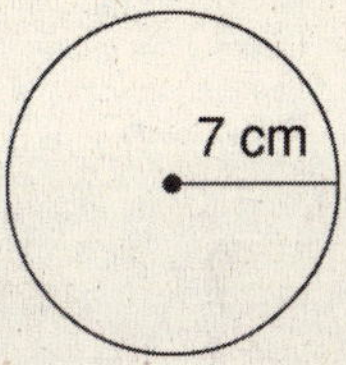

F 21.9 cm H 43.9 cm
G 22.0 cm J 44.0 cm

9.

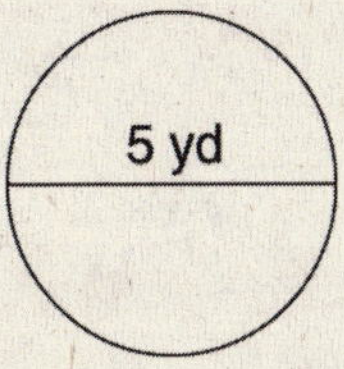

A 8.1 yd C 25.2 yd
B 15.7 yd D 31.4 yd

10.

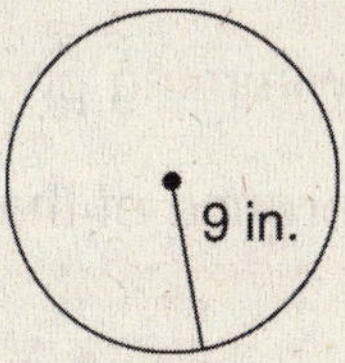

F 56.5 in. H 28.2 in.
G 28.3 in. J 14.1 in.

11. The diameter of a clock is 7 inches. What is the circumference of the clock? Use $\frac{22}{7}$ for π. ______________

Name ______________________ Date ______________ Class ______________

LESSON 9-2

Practice B

Perimeter and Circumference

Find the perimeter of each polygon.

1.

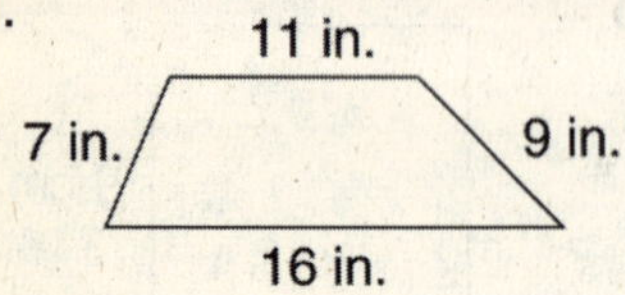

____

2.

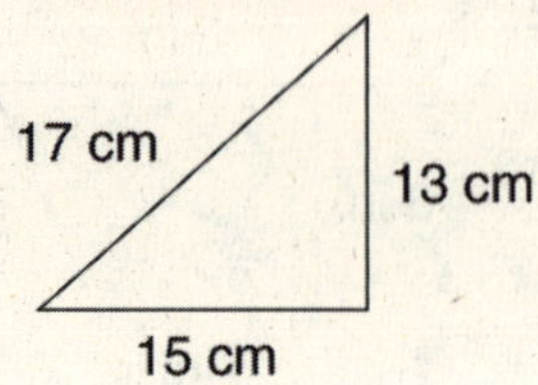

____

3.

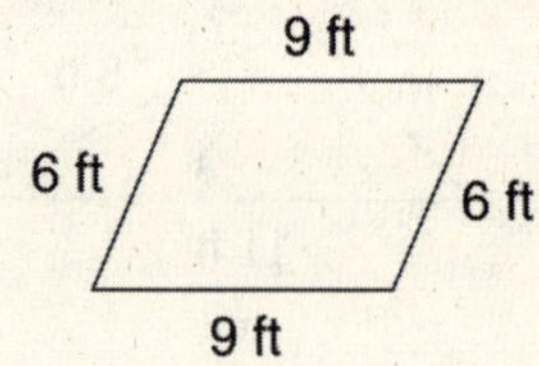

____

Find the perimeter of each rectangle.

4.

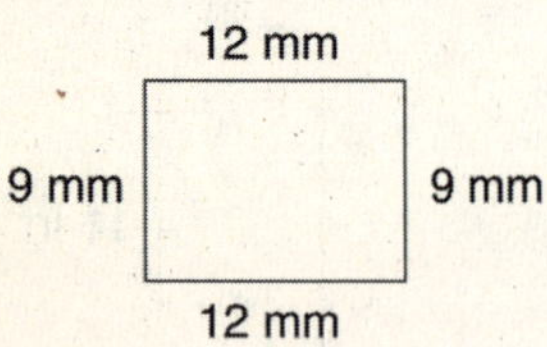

____

5.

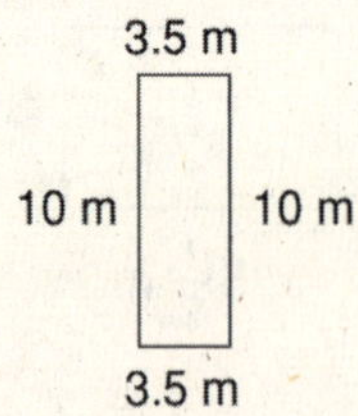

____

6.

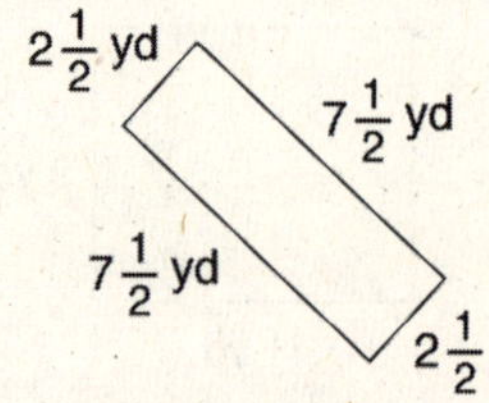

____

Find the circumference of each circle to the nearest tenth.

Use 3.14 for π or $\frac{22}{7}$.

7.

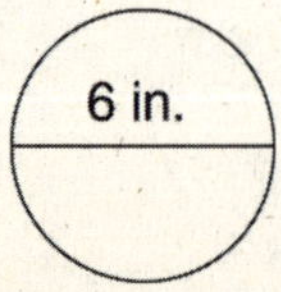

____

8.

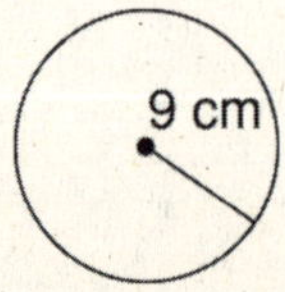

____

9.

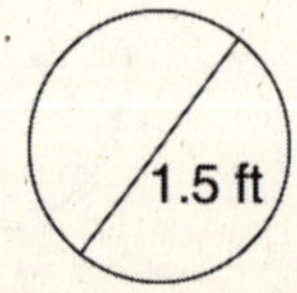

____

10. A circular swimming pool is 21 feet in diameter. What is the circumference of the swimming pool? Use $\frac{22}{7}$ for π. ______________

11. A jar lid has a diameter of 42 millimeters. What is the circumference of the lid? Use $\frac{22}{7}$ for π. ______________

12. A frying pan has a radius of 14 centimeters. What is the circumference of the frying pan? Use $\frac{22}{7}$ for π. ______________

Name ________________ Date ________________ Class ________________

LESSON 9-2

Practice C

Perimeter and Circumference

Find the perimeter of each polygon.

1.

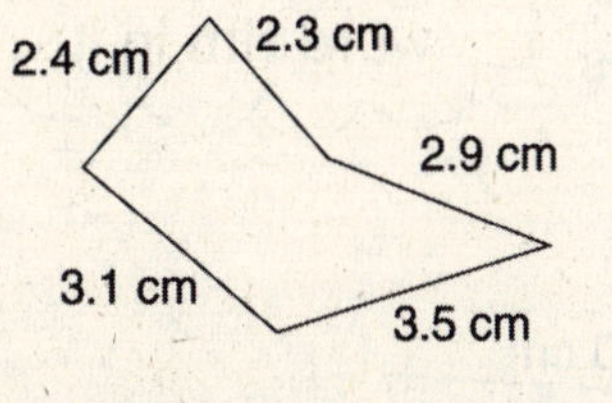

____

2.

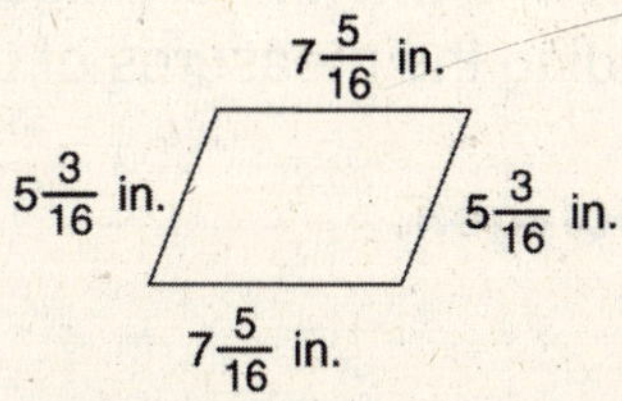

____

3.

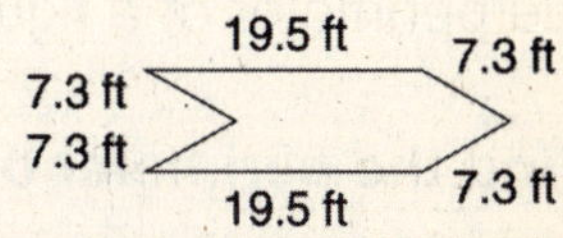

____

Find the perimeter of each rectangle.

4.

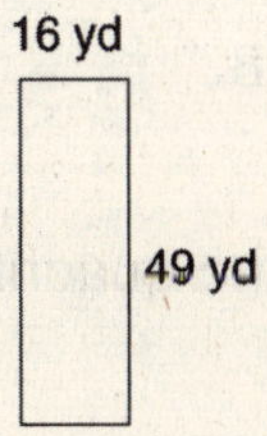

____

5.

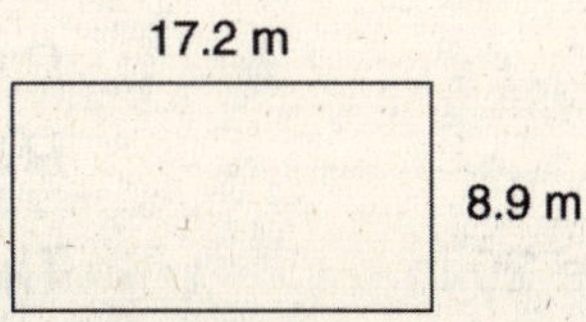

____

6.

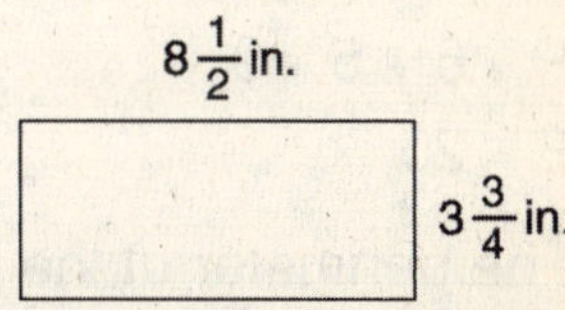

____

Find each missing measurement to the nearest tenth.

Use 3.14 for π or $\frac{22}{7}$.

7. r = 14.5 in.

d = ________________

C = ________________

8. r = ________________

d = ________________

C = 35.6 yd

9. r = ________________

d = 7.6 cm

C = ________________

10. r = ________________

d = ________________

C = 11.3 m

Solve.

11. In NCAA basketball rules, the basketball can have a maximum circumference of 30 inches. What is the maximum diameter of an NCAA basketball to the nearest hundredth? ________________

Name ________________________ Date ______________ Class ______________

LESSON 9-2

Review for Mastery

Perimeter and Circumference

The **perimeter** of any figure is the distance around the figure. Think of *perimeter* as "going around the rim" of a figure. Find the perimeter of a figure by adding the measures of the sides.

The word **perim**eter has the word **rim** in it.

Find the perimeter of each polygon.

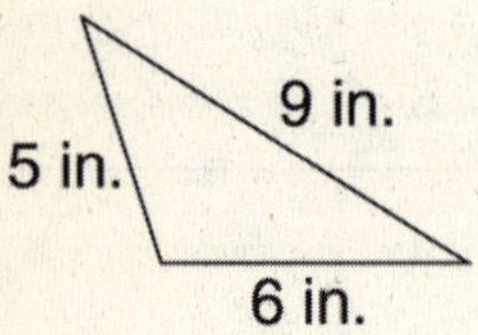

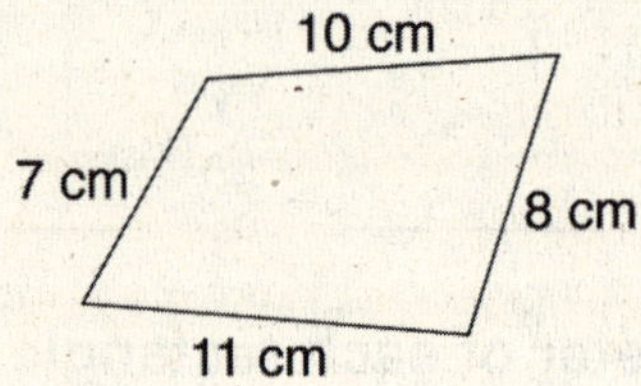

$P = 6 + 5 + 9$

$P = 20$

The perimeter of the triangle is 20 in.

$P = 7 + 11 + 10 + 8$

$P = 36$

The perimeter of the quadrilateral is 36 cm.

To find the perimeter of a rectangle, you can add the length and the width and multiply the sum by 2. The formula for the perimeter of a rectangle is $P = 2(\ell + \omega)$.

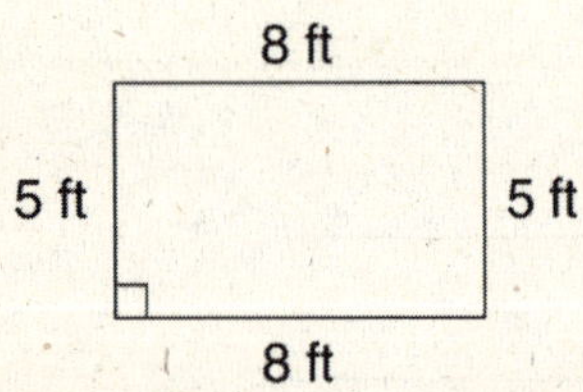

$P = 2(\ell + \omega)$

$P = 2(8 + 5)$

$P = 2(13)$

$P = 26$

The perimeter of the rectangle is 26 feet.

Find the perimeter of each polygon.

1. 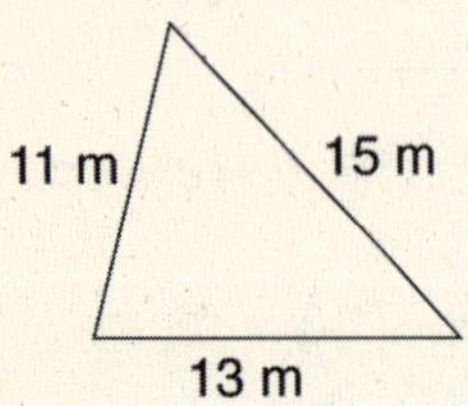

$P =$ ____ + ____ + ____

$P =$ ____ m

2. 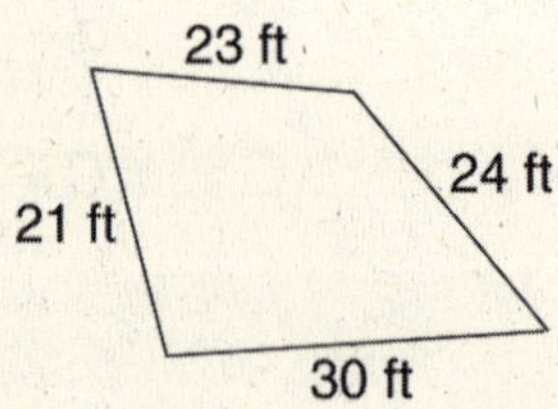

$P =$ ____ + ____ + ____ + ____

$P =$ ____ ft

3. 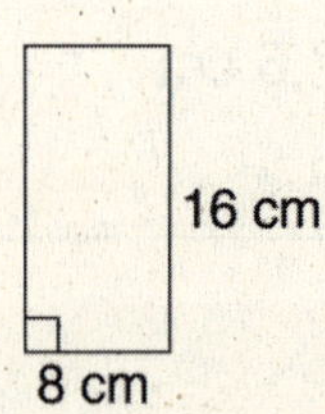

$P = 2($____ + ____$)$

$P =$ ____ cm

Name ______________________ Date ______________ Class ______________

LESSON 9-2

Review for Mastery

Perimeter and Circumference (continued)

The distance around a circle is called the **circumference.** To find the circumference of a circle, you need to know the diameter or the radius of the circle.

The ratio of the circumference of any circle to its diameter $\left(\frac{C}{d}\right)$ is always the same. This ratio is known as π (pi) and has a value of approximately 3.14.

To find the circumference C of a circle if you know the diameter d, multiply π times the diameter. $C = \pi \cdot d$, or $C \approx 3.14 \cdot d$.

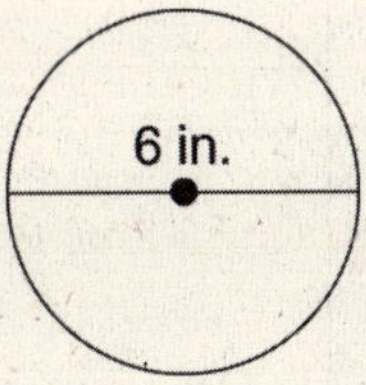

$C = \pi \cdot d$

$C \approx 3.14 \cdot d$

$C \approx 3.14 \cdot 6$

$C \approx 18.84$

The circumference is about 18.8 in. to the nearest tenth.

The diameter of a circle is twice as long as the radius r, or $d = 2r$. To find the circumference if you know the radius, replace d with $2r$ in the formula. $C = \pi \cdot d = \pi \cdot 2r$

Find the circumference given the diameter.

4. $d = 9$ cm

$C = \pi \cdot d$

$C \approx 3.14 \cdot$ ______

$C \approx$ ________

The circumference is ______ cm to the nearest tenth of a centimeter.

Find the circumference given the radius.

5. $r = 13$ in.

$C = \pi \cdot 2r$

$C \approx 3.14 \cdot (2 \cdot$ ______$)$

$C \approx 3.14 \cdot$ ______

$C \approx$ ________

The circumference is ______ in. to the nearest tenth of an inch.

Find the circumference of each circle to the nearest tenth. Use 3.14 for π.

6.

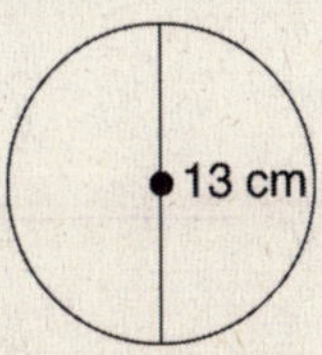

____

7.

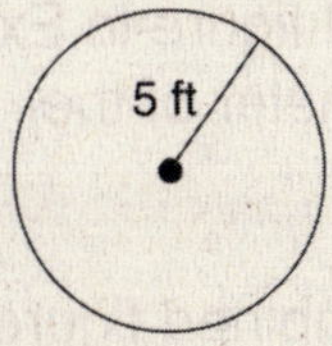

____

8.

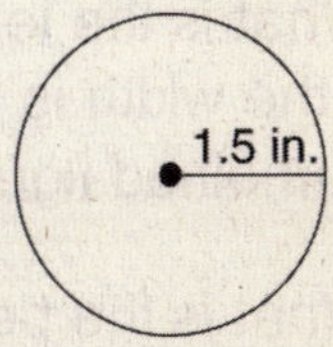

____

Name ______________________ Date ______________ Class ______________

LESSON 9-2

Challenge

All-Around Formulas

Use the first figure in each row to write a formula for the perimeter of the combined figure next to it. Use your formulas in Exercises 5–8.

	Original Figure	Combined Figure	Perimeter
1.	A regular octagon: 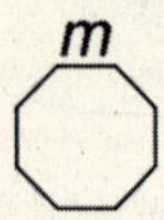	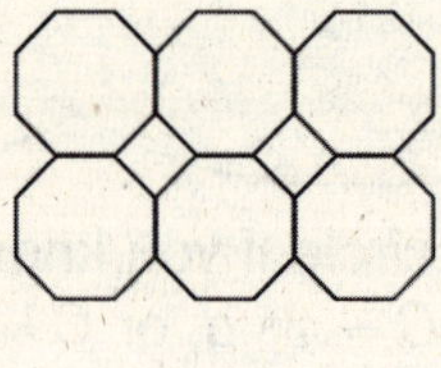	
2.	An isosceles triangle: 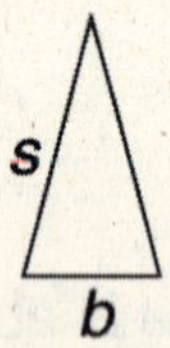	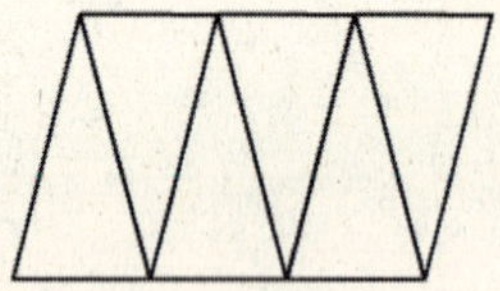	
3.	A parallelogram: 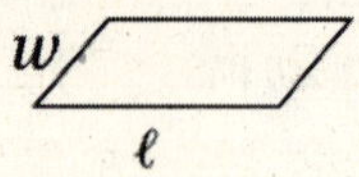	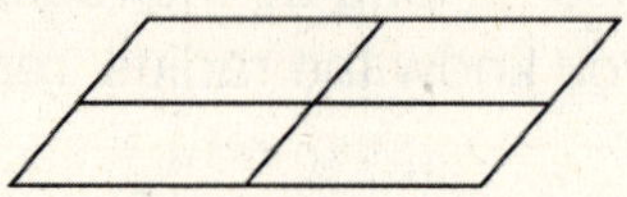	
4.	A semicircle: 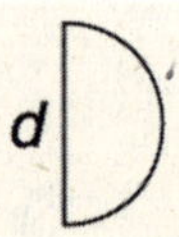	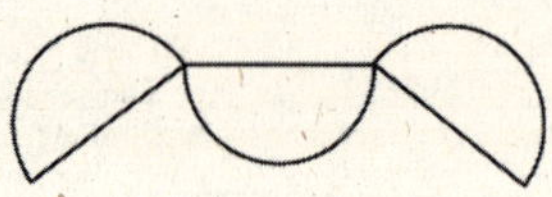	

5. What is the perimeter of the combined figure in Exercise 1 if $m = 4$ in.? ____________

6. What is the perimeter of the combined figure in Exercise 2 if $s = 4.5$ m and $b = 5.2$ m? ____________

7. What is the length of the original figure in Exercise 3 if the width is 6 in. and the perimeter of the combined figure is 56 in.? ____________

8. What is the perimeter of the combined figure in Exercise 4 if $d = 10$ cm? ____________

Name ______________________ Date ______________ Class ______________

LESSON 9-2

Problem Solving

Perimeter and Circumference

Write the correct answer.

1. Mr. Marcos, the gym teacher, had the seventh graders run around the perimeter of the gym 3 times. The gym has a length of 34 feet and a width of 58 feet. What was the total distance the students ran?

____

2. The distance between bases on a baseball field is 90 feet. If 3 players hit home runs during a game and each runs around all 4 bases, what is the total distance the players run?

____

3. Basketball rims have a diameter of 18 inches. If you want to put a band around a basketball rim, how much material to the nearest tenth of an inch will you need?

____

4. A pizza cutter has a diameter of 2.5 inches. To cut a pizza in half, the cutter makes two complete revolutions. What is the diameter of the pizza?

____

5. A round stained-glass window has a circumference of 195 inches. What is the radius of the window to the nearest inch?

____

6. A planter full of pansies has a diameter of 14 inches. What is the circumference of the planter to the nearest inch?

____

Choose the letter of the correct answer.

7. A welcome mat on the front porch is a semicircle. The straight side of the mat is 36 inches. What is the perimeter of the mat?

   A 92.52 in. C 56.52 in.
   B 64.26 in. D 28.26 in.

8. The radius of the planet Jupiter is about 44,368 miles. What is the approximate circumference of Jupiter to the nearest mile?

   F 557,262 mi H 139,316 mi
   G 278,631 mi J 69,658 mi

9. Four square tables with sides of 48 inches each are placed end to end to form one big table. What is the perimeter of the table that is formed?

   A 192 in. C 480 in.
   B 384 in. D 768 in.

10. The Great Pyramid at Giza, Egypt, has a base with four equal sides. If the perimeter of the pyramid is 3,023 feet, what is the length (to the nearest foot) of one side of the base of the pyramid?

    F 754 ft H 756 ft
    G 755 ft J 757 ft

Name ______________________ Date ______________ Class ______________

LESSON 9-2

Reading Strategies

Use A Graphic Organizer

Perimeter is the distance around a polygon.

Circumference is the distance around a circle.

The chart below shows formulas for finding the perimeter of polygons and the circumference of circles.

Measuring the Distance Around Figures

Perimeter of Polygons
Add lengths of sides.

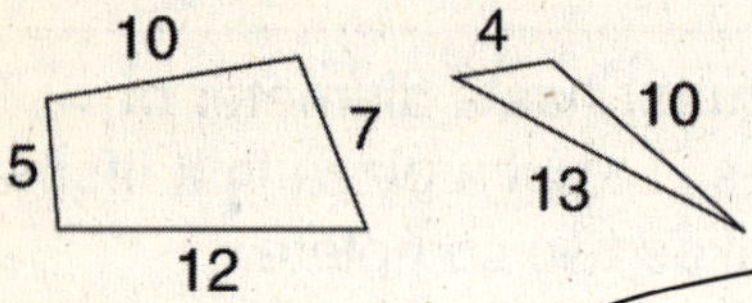

Perimeter of Rectangles

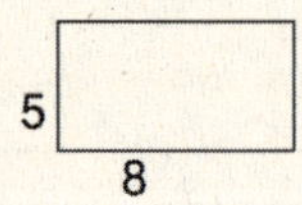

$P = 2\ell + 2\omega$

(Perimeter = twice the length plus twice the width)

Circumference of Circle

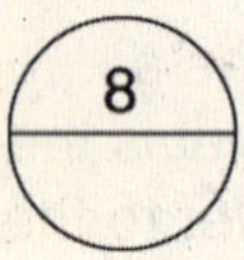

$C = \pi d$

(Circumference = pi • diameter)

Circumference of Circle

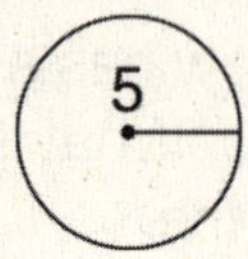

$C = 2\pi r$

(Circumference = 2 • pi • radius)

Use the information in the chart above to complete each exercise.

1. Write the side lengths for the triangle shown above.

____

2. How would you find the perimeter of the triangle?

____

3. If you knew the radius of a circle, what formula would you use to find its circumference?

____

4. Write the formula for finding the perimeter of a rectangle.

____

5. If you knew the diameter of a circle, what formula would you use to find its circumference?

____

Name ______________________ Date ______________ Class ______________

LESSON 9-2

Puzzles, Twisters & Teasers

Around and Around!

Find the perimeter or circumference of each figure. Round the circumference to the nearest tenth. Match your answers to the corresponding letter to solve the riddle.

1. ______________ = **E**

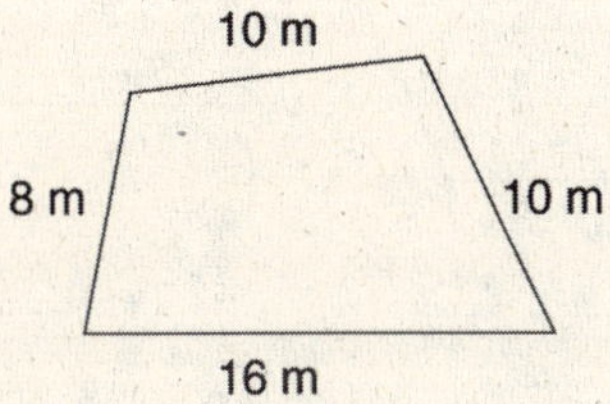

2. ______________ = **V**

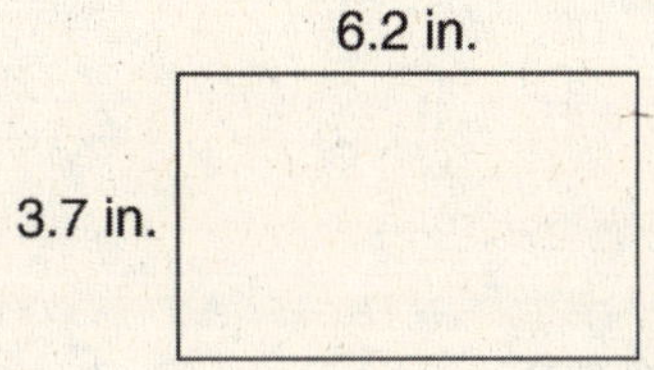

3. ______________ = **L**

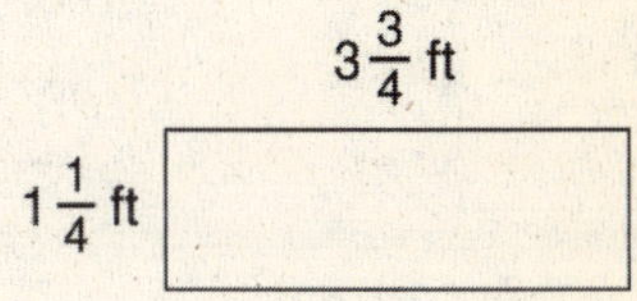

4. ______________ = **I**

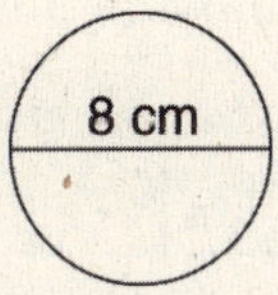

5. ______________ = **C**

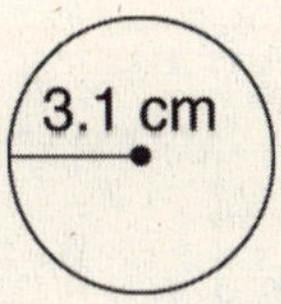

6. ______________ = **O**

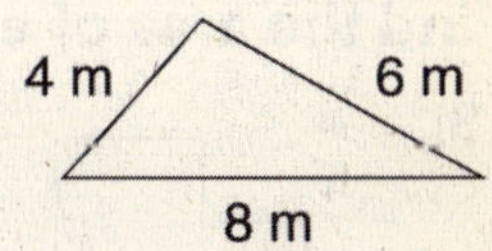

7. ______________ = **R**

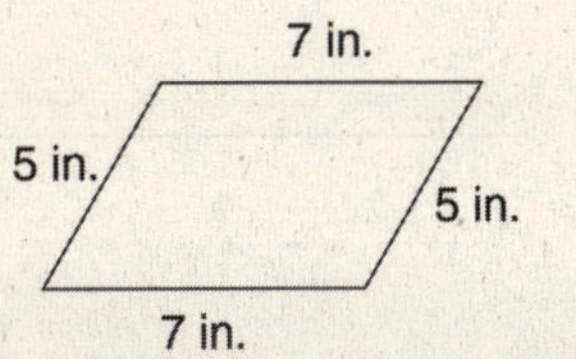

8. ______________ = **S**

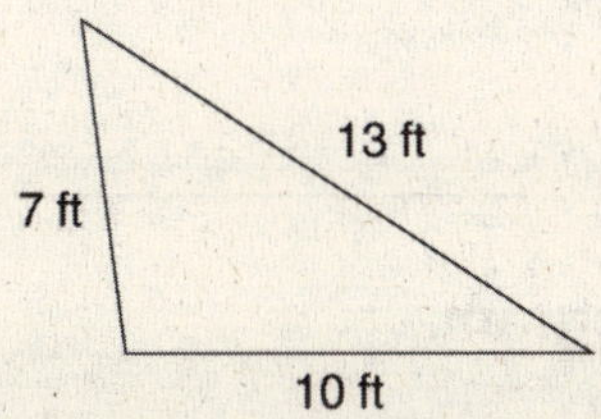

9. ______________ = **U**

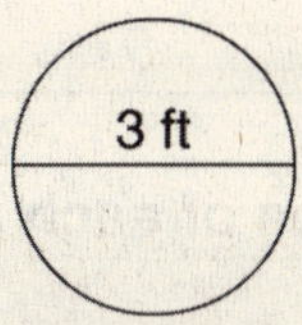

What's round and bad-tempered?

A ___ ___ ___ ___ ___ ___ ___
19.8 25.1 19.5 25.1 18 9.4 30

___ ___ ___ ___ ___ ___
19.5 25.1 24 19.5 10 44

Name ______________________ Date ______________ Class ______________

LESSON 9-3

Practice A

Area of Parallelograms

Find the area of each figure. Choose the letter for the best answer.

1.

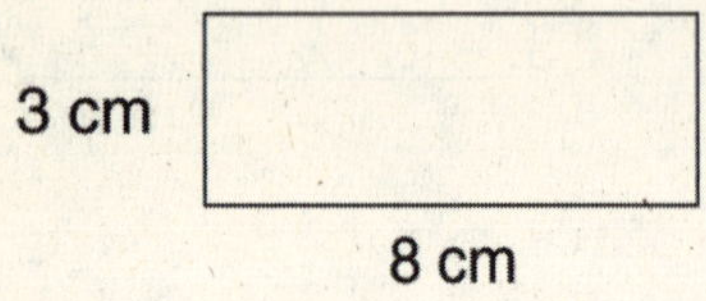

2.

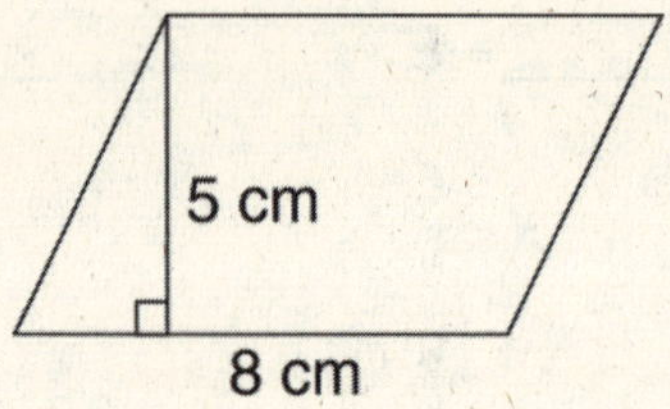

A 11 cm^2 C 24 cm^2

B 22 cm^2 D 48 cm^2

F 26 cm^2 H 48 cm^2

G 40 cm^2 J 80 cm^2

Find the area of each rectangle.

3.

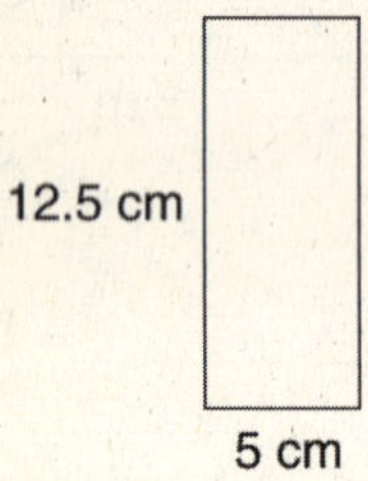

4.

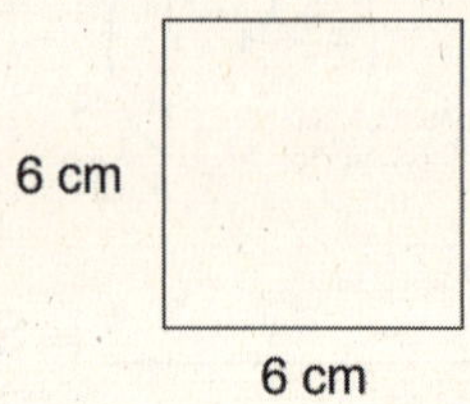

5.

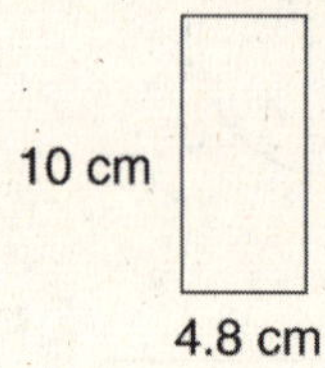

______________ ______________ ______________

Find the area of each parallelogram.

6.

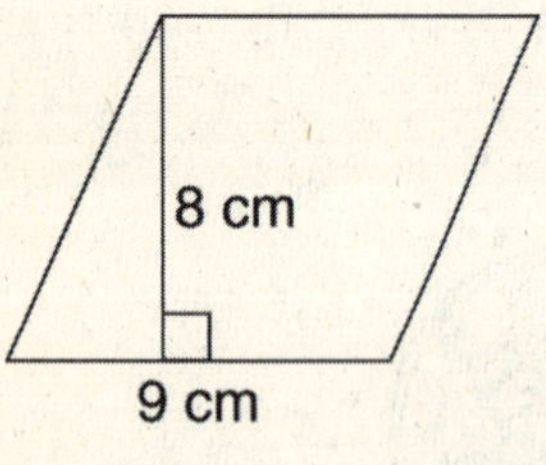

7.

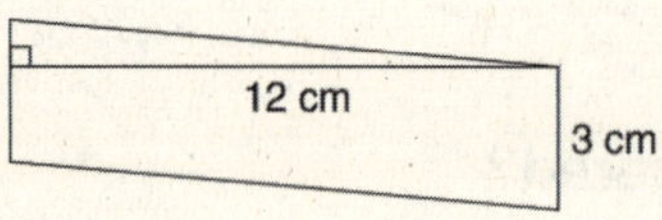

8.

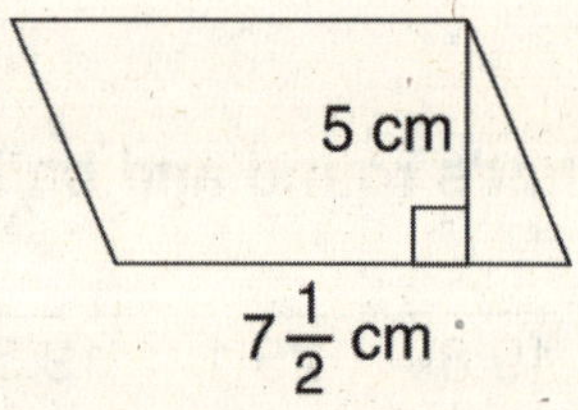

______________ ______________ ______________

9. Michelle wants to carpet her rectangular living room floor. The area of the living room is 192 ft^2. The length of the living room is 16 ft. What is the width of the living room? ______________

10. Mustafa is tiling his bathroom. The section that needs to be tiled is 62 in. by 70 in. How many square inches of tile does he need? ______________

Name ______________________ Date ______________ Class ______________

LESSON 9-3

Practice B

Area of Parallelograms

Find the area of each rectangle.

1.
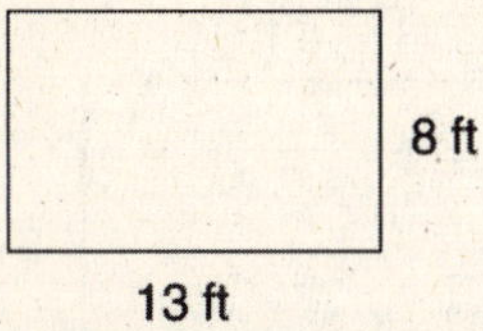

____

2.
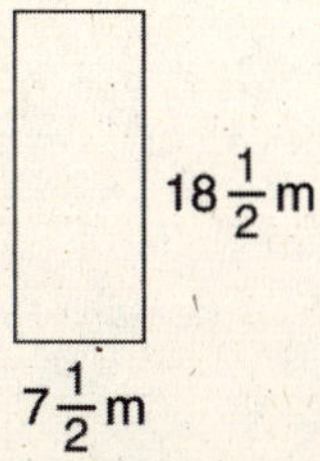

____

3.
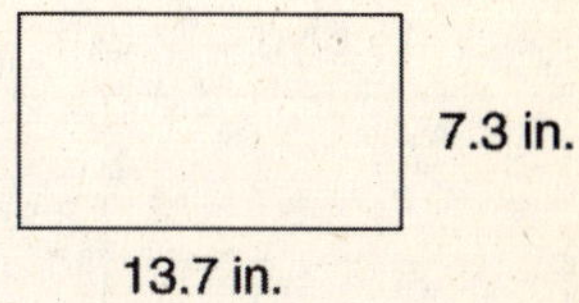

____

Find the area of each parallelogram.

4.
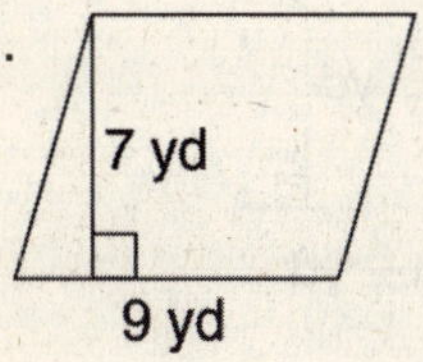

____

5
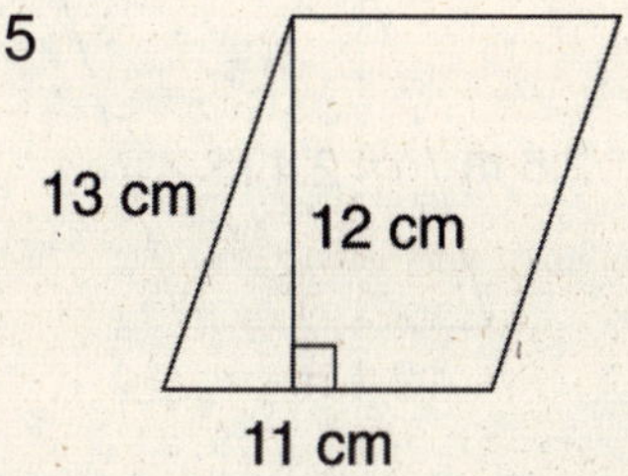

____

6.
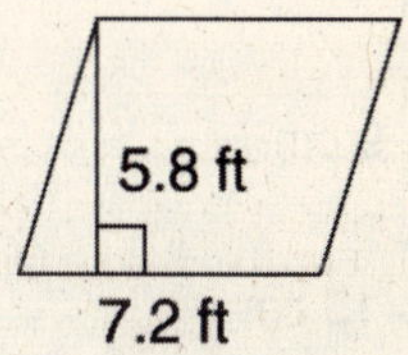

____

7.
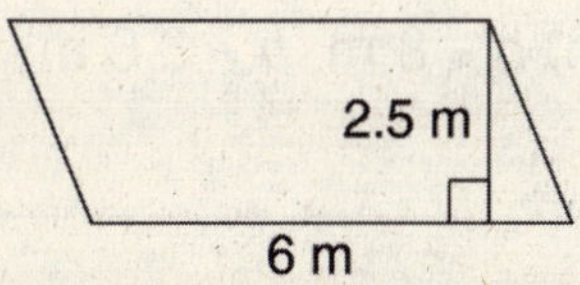

____

8.
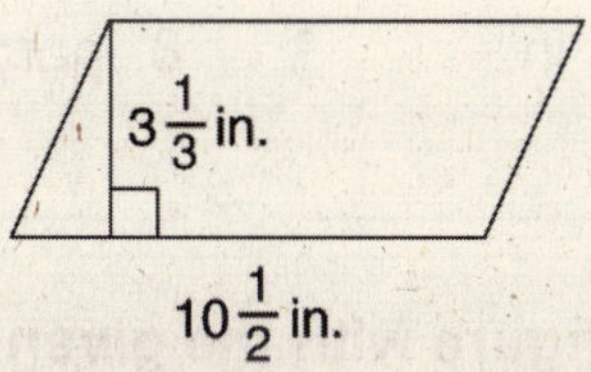

____

9.
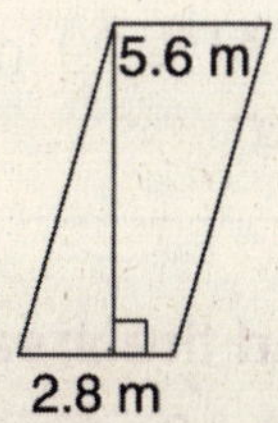

____

10. A dollar bill is 15.5 cm long and 6.5 cm wide. What is the area of a dollar bill?

____

11. A rectangular hallway has an area of 70 ft^2. The width of the hallway is 4 feet. What is the length of the hallway?

____

Name ______________________ Date ______________ Class ______________

LESSON 9-3

Practice C

Area of Parallelograms

Find the area of each rectangle.

1.
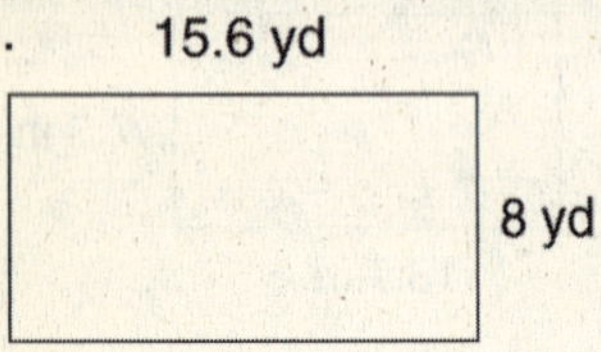

____

2.
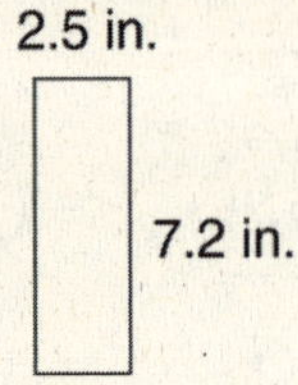

____

3.
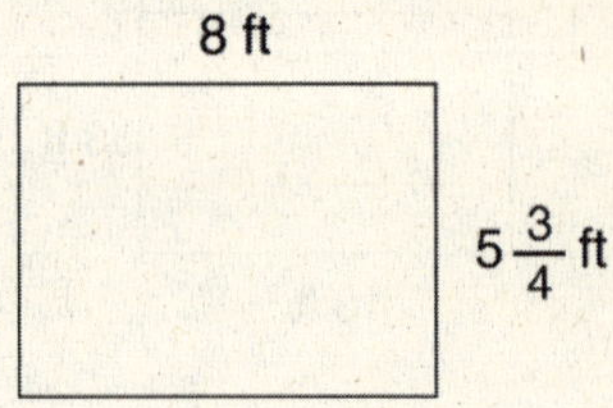

____

Find the area of each parallelogram.

4.
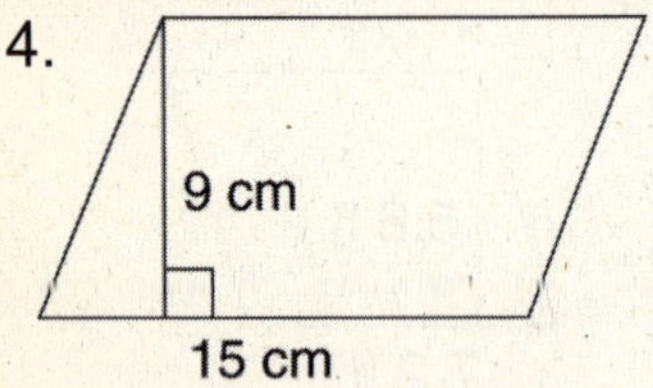

____

5.
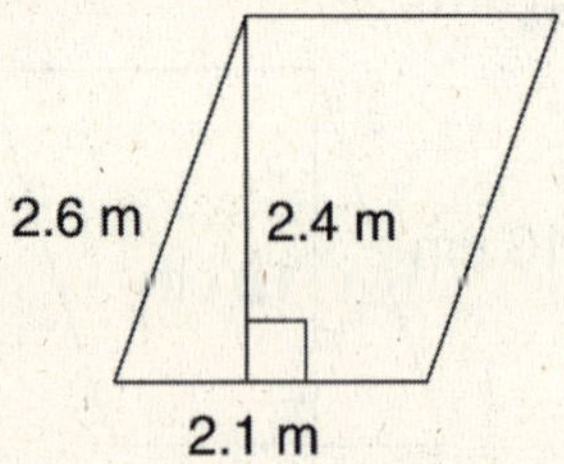

____

6.
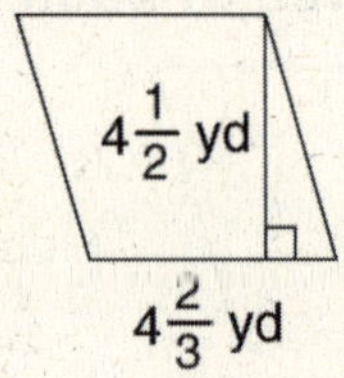

____

Find the area of each polygon.

7. rectangle: $\ell = 29$ in., $\omega = 15$ in.

____

8. parallelogram: $b = 8$ m, $h = 3.6$ m

____

For 9–12, find the area of the figure with the given vertices.

9. (3, –2), (8, –2), (3, 6), (8, 6)

____

10. (2, 0), (5, 3), (8, 0), (11, 3)

____

11. (–4, 1), (–9, 1), (0, 7), (–5, 7)

____

12. (–1, –2), (4, –2), (–1, 5), (4, 5)

____

13. What is the length of the base of a parallelogram with an area of 36 cm^2 and a height of 4.5 cm?

____

14. A window is composed of 6 windowpanes. Each has a length of $10\frac{1}{2}$ in. and a width of $7\frac{1}{4}$ in. What is the total area of glass in the window?

____

Name ______________________ Date ______________ Class ______________

LESSON 9-3

Review for Mastery

Area of Parallelograms

The **area** of a figure is the number of square units inside the figure.

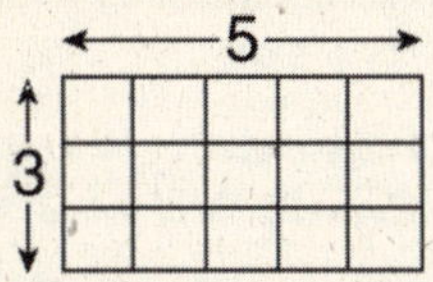

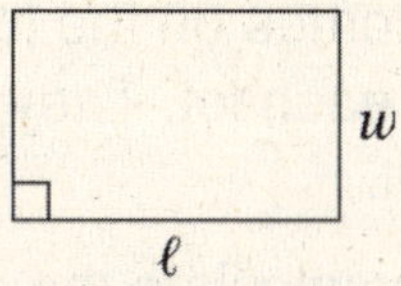

You can count the squares inside the rectangle. There are 15 square units within the rectangle. This is equal to 5 • 3.

To find the area of a rectangle, multiply the length (ℓ) times the width (ω).

$$A = \ell \cdot \omega$$

Find the area of each rectangle.

1.

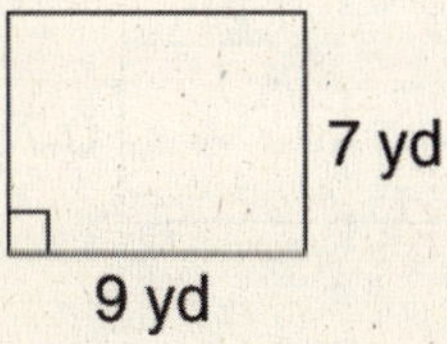

$A = l \cdot w$

$A =$ _____ • _____

$A =$ _____

The area is _____ yd^2.

2.

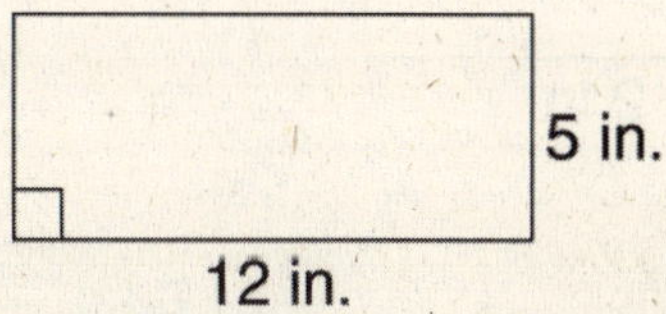

$A = l \cdot w$

$A =$ _____ • _____

$A =$ _____

The area is _____ in^2.

To find the area of a parallelogram, multiply the base b times the height h.

$$A = b \cdot h$$

Find the area of each parallelogram.

3.

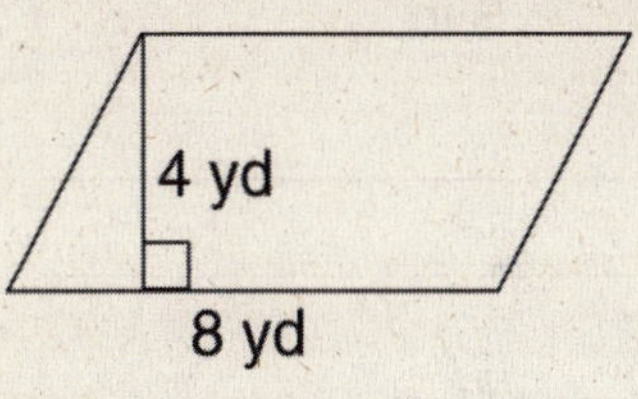

$A = b \cdot h$

$A =$ _____ • _____

$A =$ _____

The area is _____ yd^2.

4.

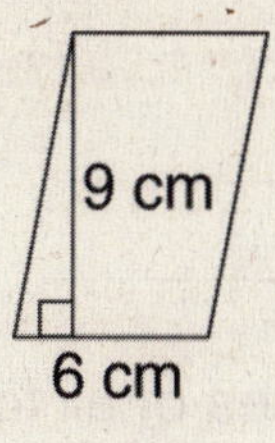

$A = b \cdot h$

$A =$ _____ • _____

$A =$ _____

The area is _____ cm^2.

Name ______________________ Date ______________ Class ______________

LESSON 9-3

Challenge

Size It Up!

Gonzalez Builders purchased the four lots shown below. The company intends to build homes on the lots and needs to know the total area in square feet of each lot. They also need to know the perimeter in feet of each lot.

You can use proportions to calculate the areas and perimeters. For each figure, the scale is 1 inch: 60 feet.

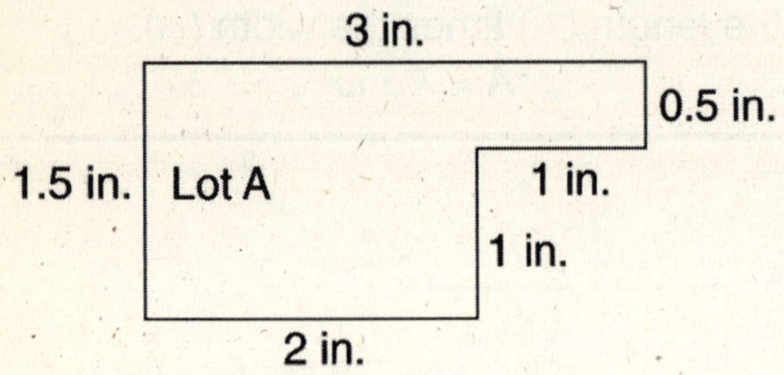

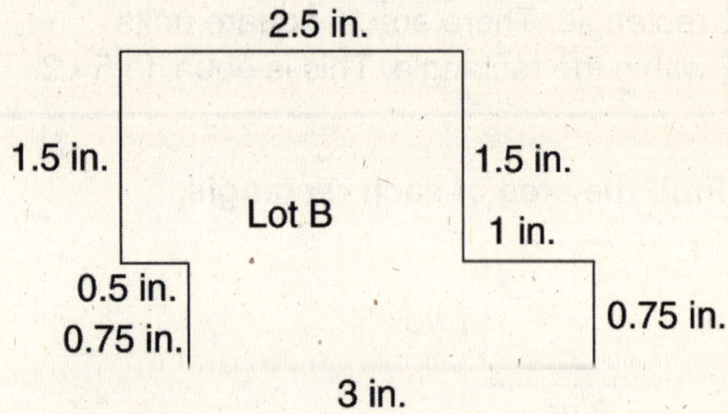

1. total area ______________________
   perimeter ______________________

2. total area ______________________
   perimeter ______________________

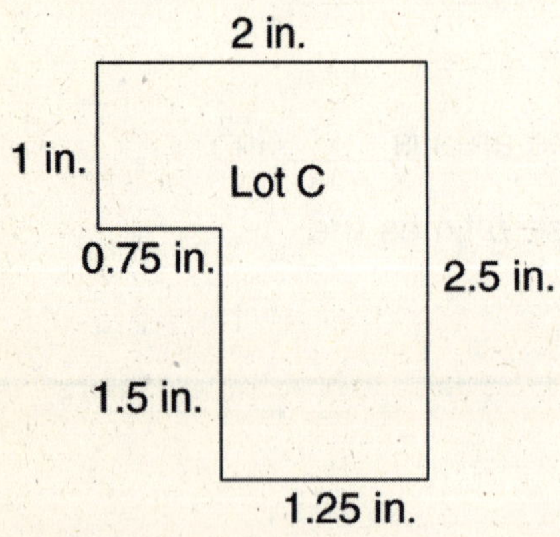

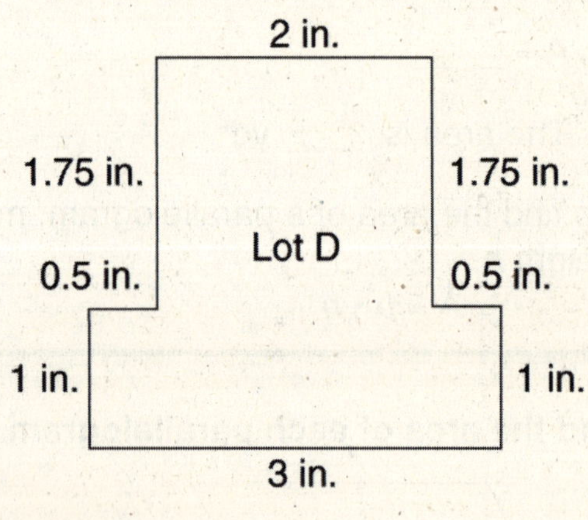

3. total area ______________________
   perimeter ______________________

4. total area ______________________
   perimeter ______________________

5. What is the combined area of all four lots? ______________________

6. What length of fencing would be needed to enclose all four lots during construction? ______________________

Name ______________________ Date ______________ Class ______________

LESSON 9-3

Problem Solving

Area of Parallelograms

Write the correct answer.

1. A dollar bill has an area of 15.86 square inches. If a dollar bill is 2.6 inches long, how wide is it?

   ____

2. On an official United States flag, the ratio of width to length is exactly 1 to 1.9. What is the area of a United States flag whose width is 2 feet?

   ____

3. A back yard is shaped like a parallelogram with a height of 32 feet and a base of 100 feet. One bag of grass seed covers 125 square feet. What is the least number of bags of seed needed to seed the lawn?

   ____

4. The art club is painting a mural on a school wall. The mural is in the shape of a parallelogram. If the base of the mural is 10.5 feet long and the mural covers 89.25 square feet, how high is the mural?

   ____

Choose the letter of the correct answer.

5. In baseball, the area of each base is 225 square inches. Each base is a square. What is the length of each side of a base on a baseball field?

   A 12 in.
   C 25 in.
   B 22.5 in.
   D 15 in.

6. The area of a parallelogram is 632.1 square centimeters. Its base is 24.5 centimeters. What is the height of the parallelogram?

   F 25.8 cm
   H 21.9 cm
   G 705.6 cm
   J 11.8 cm

7. The official rules for volleyball were developed in 1897. The rules state that the court or floor space must be 25 feet wide and 50 feet long. An official basketball court is 94 feet by 50 feet. How much larger is the area of a basketball court than the area of a volleyball court?

   A 69 ft^2 larger
   B 3,450 ft^2 larger
   C 1,250 ft^2 larger
   D 4,700 ft^2 larger

8. Two parallelograms each have an area of 288 square inches. One has a height of 12 inches, and the other has a height of 18 inches. What are the bases of each parallelogram?

   F 40 in. and 30 in.
   G 22 in. and 15 in.
   H 24 in. and 16 in.
   J 26 in. and 20 in.

Name ______________________ Date ______________ Class ______________

LESSON 9-3

Reading Strategies

Compare and Contrast

Area measures the surface a figure covers.
Area is measured in square units.

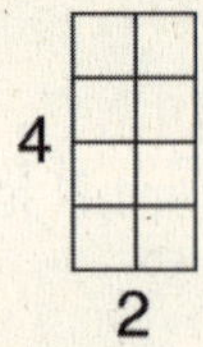

To find the area of this rectangle, count the number of square units. There are eight square units.

Another way to find the area is to use this formula: $A = \ell\omega$.
Read: "Area equals length times width."

2 units times 4 units = 8 square units.
Area = 8 square units

Compare a parallelogram to a rectangle. The width of a parallelogram is called the height. The length of a parallelogram is called the base.

Moving one piece of the parallelogram shows how the base and height compare to the length and width of a rectangle.

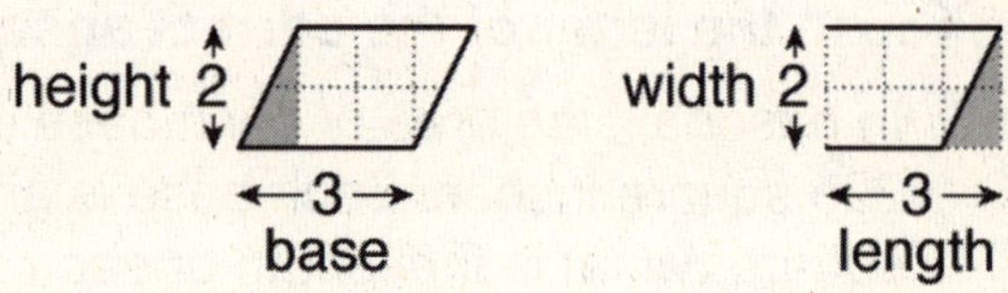

Area = base • height ($A = bh$)	Area = length • width ($A = \ell\omega$)
$A = 3 \cdot 2$	$A = 3 \cdot 2$
$A = 6$ square units	$A = 6$ square units

1. The height of a parallelogram is the same as what part of a rectangle?

____

2. The base of a parallelogram is the same as what part of a rectangle?

____

3. What is the same about finding the area of a parallelogram and a rectangle?

____

____

4. What is different about finding the area of a parallelogram and a rectangle?

____

Name ______________________ Date ____________ Class ____________

LESSON 9-3

Puzzles, Twisters & Teasers

Reach the Heights!

Across

4. a four-sided figure with congruent and parallel opposite sides
5. shape
8. the width of a parallelogram

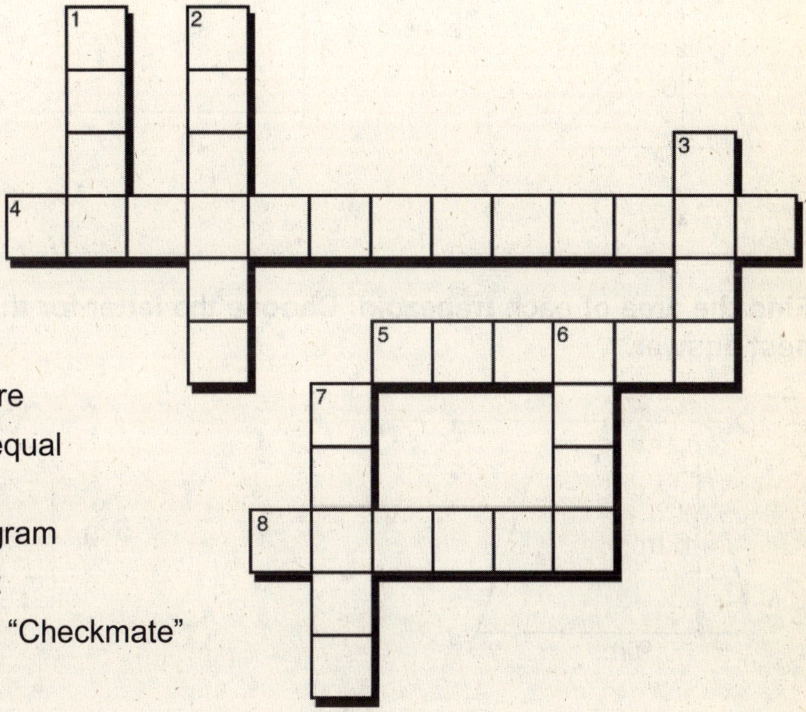

Down

1. number of square units needed to cover the figure
2. parallelogram with four equal sides and angles
3. the length of a parallelogram
6. ______ of measurement
7. a ______ board leads to "Checkmate"

Name ______________________ Date ______________ Class ______________

LESSON 9-4

Practice A

Area of Triangles and Trapezoids

Find the area of each triangle.

1. 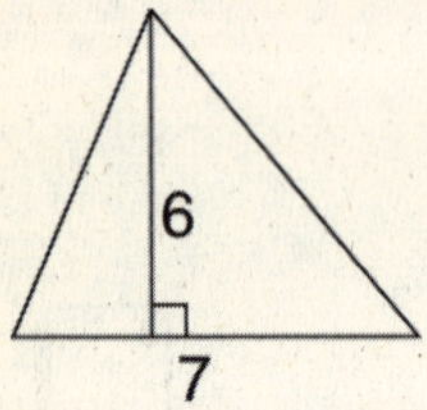

____

2. 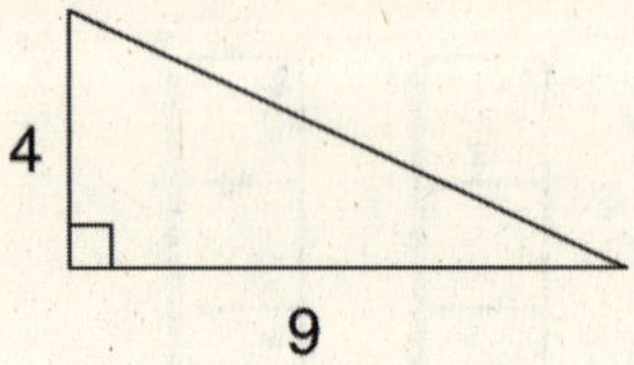

____

3. 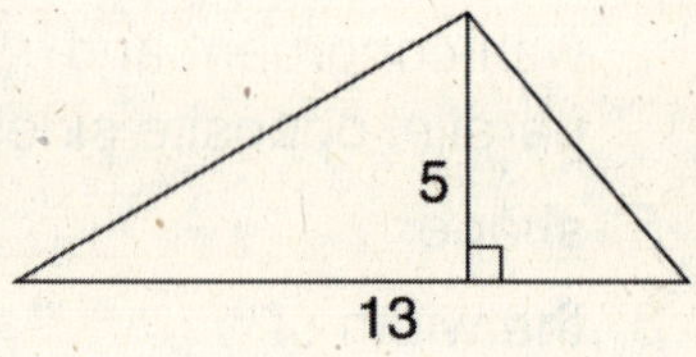

____

Find the area of each trapezoid. Choose the letter for the best answer.

4. 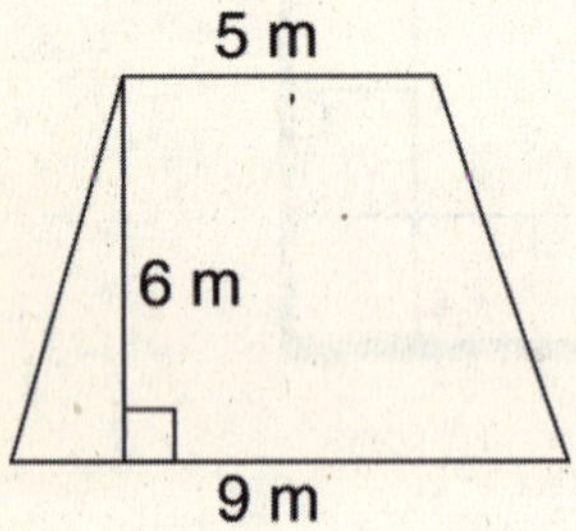

A $24\ m^2$
B $42\ m^2$
C $54\ m^2$
D $84\ m^2$

5. 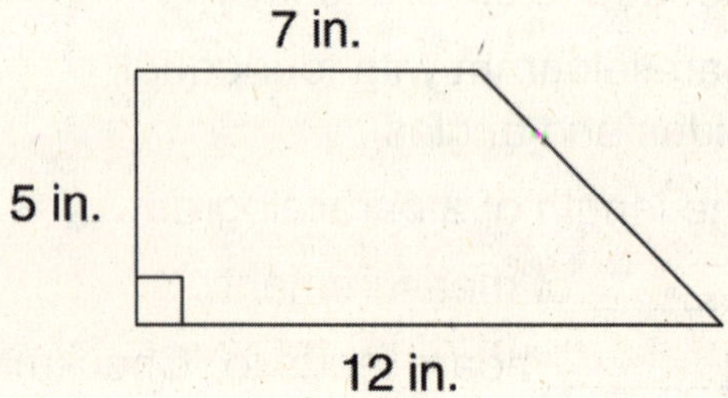

F $95\ in^2$
G $60\ in^2$
H $47.5\ in^2$
J $25\ in^2$

6. 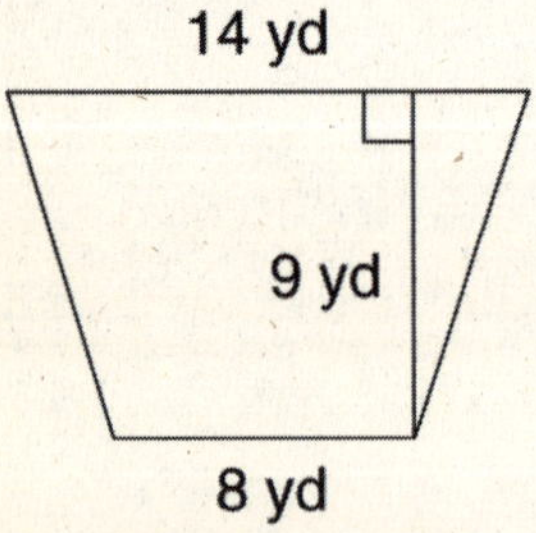

A $99\ yd^2$
B $126\ yd^2$
C $198\ yd^2$
D Not Here

7. 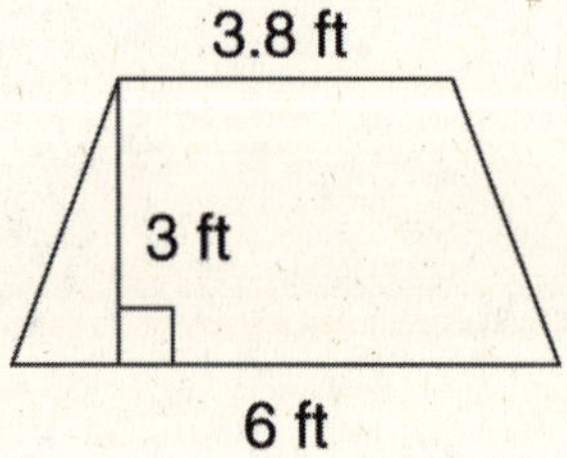

F $11.4\ ft^2$
G $14.7\ ft^2$
H $18\ ft^2$
J $29.4\ ft^2$

8. The state of Missouri is shaped somewhat like a trapezoid. What is the approximate area of Missouri?

____

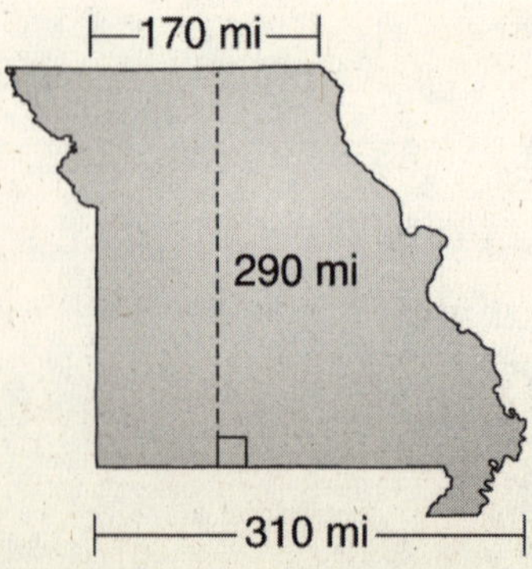

Name ____________________ Date ____________ Class ____________

LESSON 9-4

Practice B

Area of Triangles and Trapezoids

Find the area of each triangle.

1.

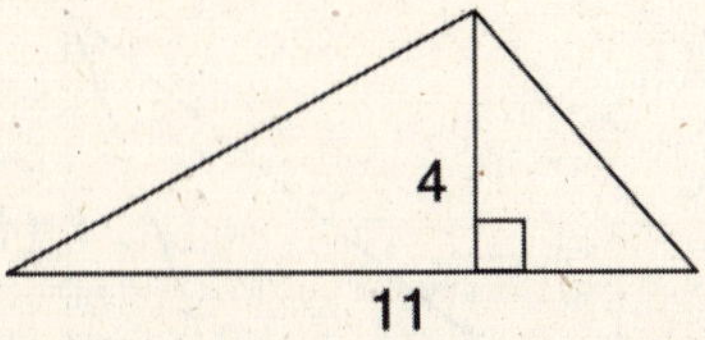

____

2.

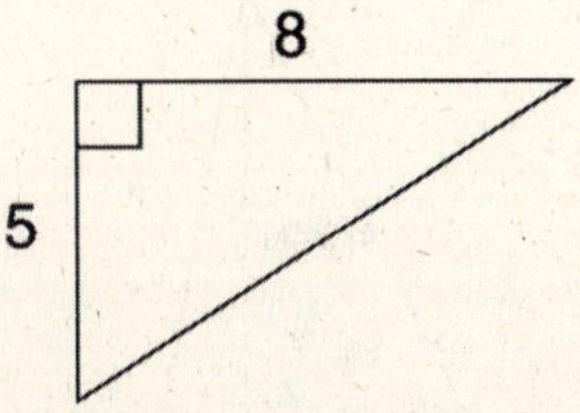

____

3.

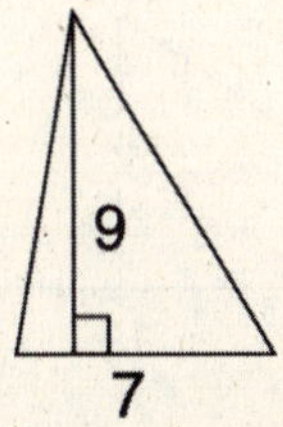

____

4.

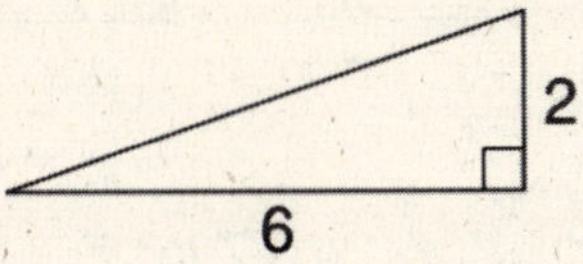

____

5.

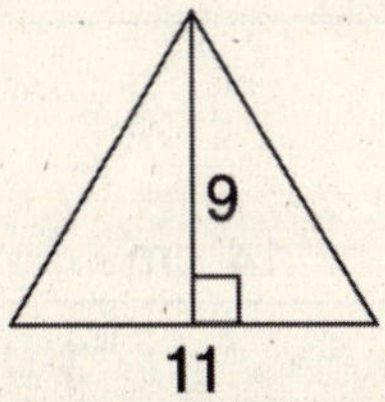

____

6.

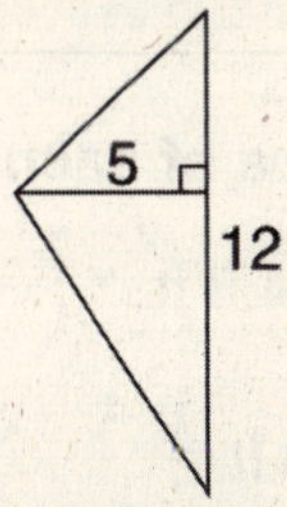

____

Find the area of each trapezoid.

7.

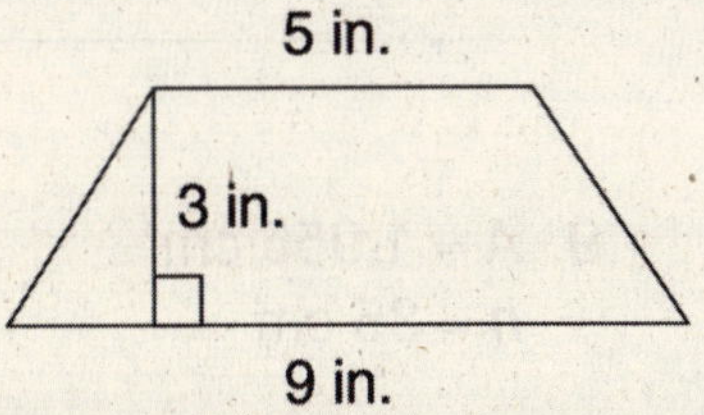

____

8.

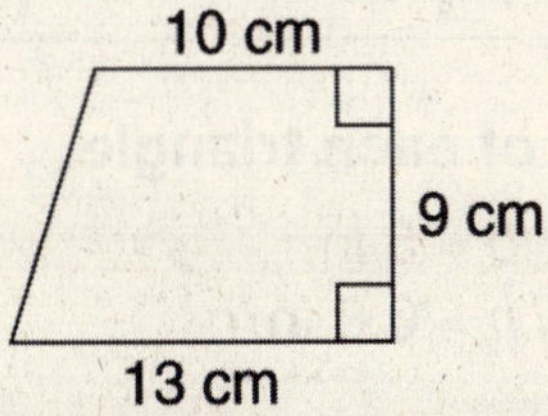

____

9.

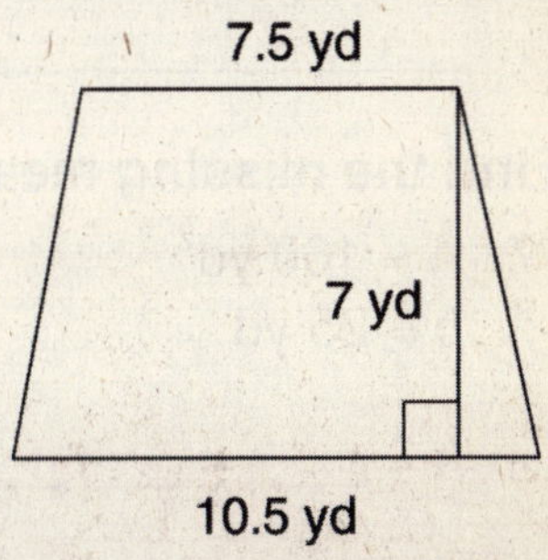

____

10.

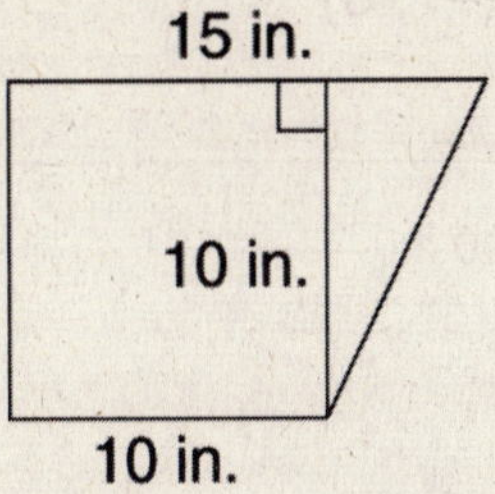

____

11.

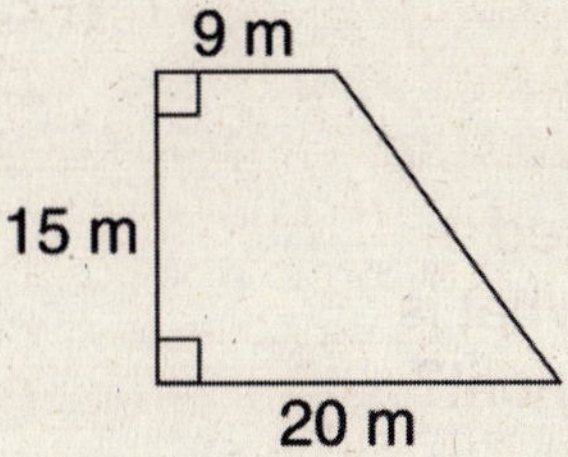

____

12.

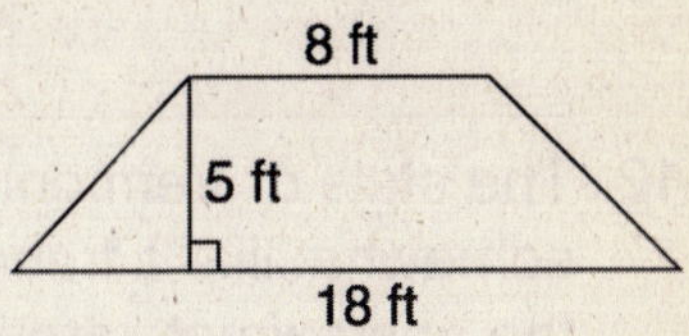

____

13. The state of Montana is shaped somewhat like a trapezoid. What is the approximate area of Montana?

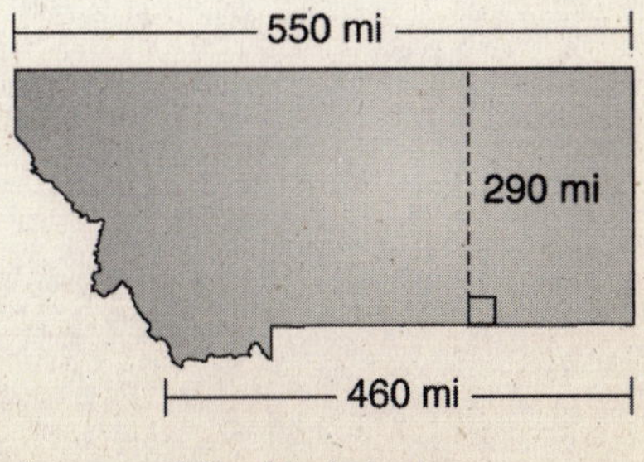

____

Name ______________________ Date ______________ Class ______________

LESSON 9-4

Practice C

Area of Triangles and Trapezoids

Find the area of each triangle.

1.

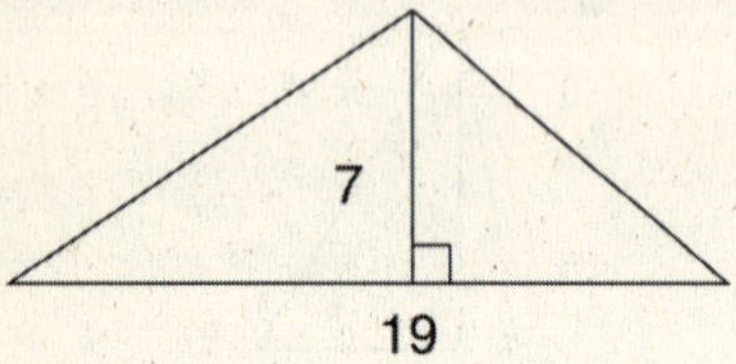

____

2.

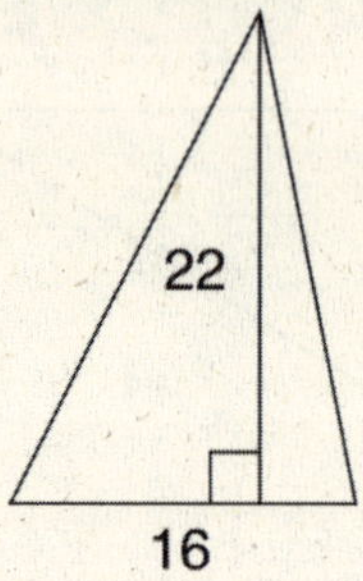

____

3.

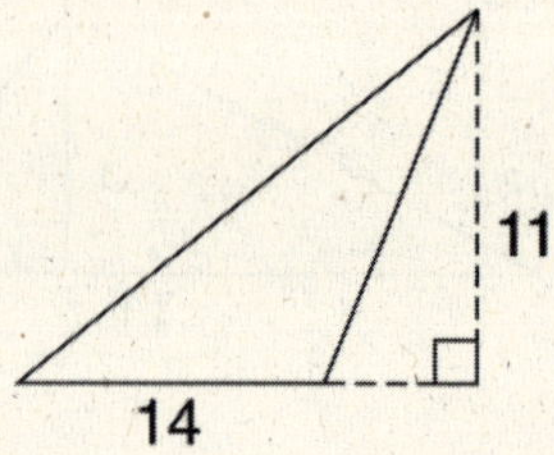

____

Find the area of each trapezoid.

4.

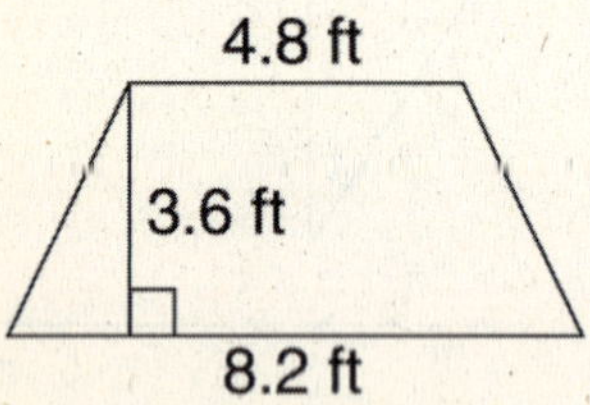

____

5.

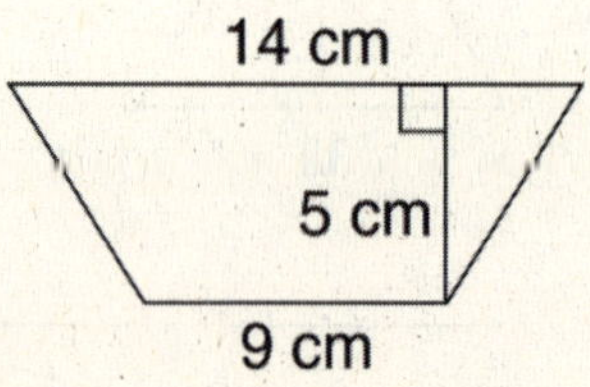

____

6.

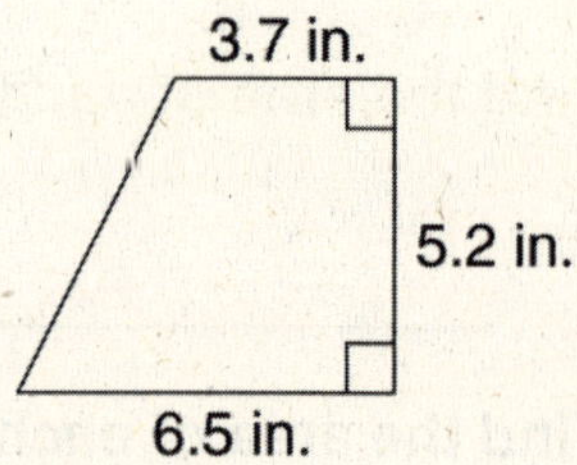

____

Find the missing measurement of each triangle.

7. $A = 100 \text{ yd}^2$
$b = 25$ yd

$h =$ ________________

8. $b = 5$ in.
$h = 0.8$ in.

$A =$ ________________

9. $A = 1{,}955 \text{ cm}^2$
$h = 85$ cm

$b =$ ________________

For 10–11, find the area of the polygon with the given vertices.

10. (2, 3), (5, 7), (10, 3), (9, 7)

____

11. (–2, 6), (–7, 1), (–7, 6)

____

12. The state of Vermont is shaped somewhat like a trapezoid. What is the approximate area of Vermont?

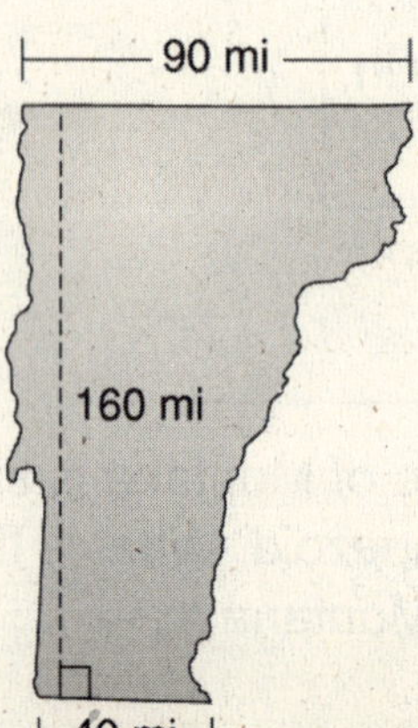

____

Name ______________________ Date ______________ Class ______________

LESSON 9-4

Review for Mastery

Area of Triangles and Trapezoids

The diagram shows how you can cut a parallelogram into two congruent triangles.

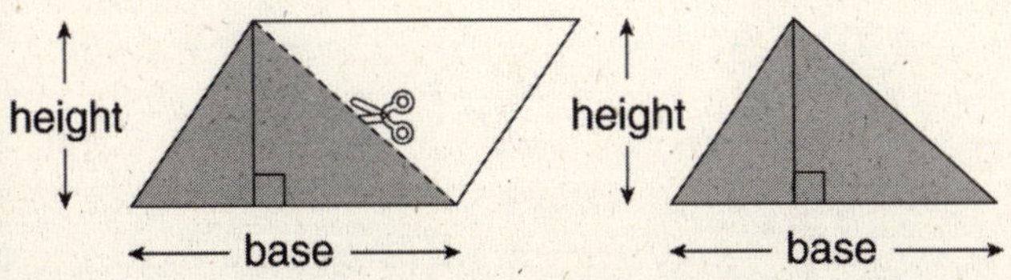

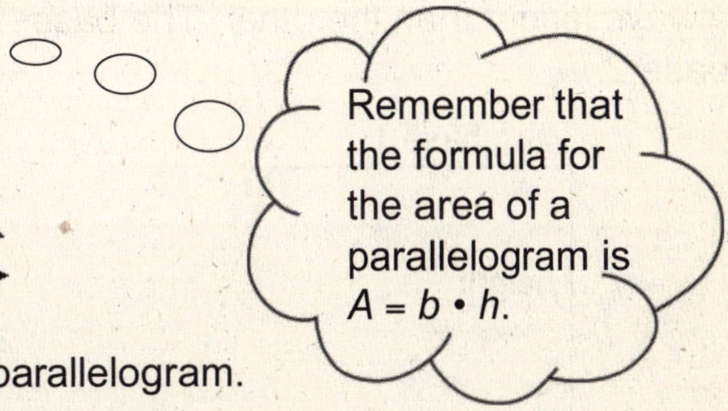

The area of the triangle is $\frac{1}{2}$ the area of the parallelogram.

The formula for the area of a triangle is $A = \frac{1}{2} \cdot b \cdot h$.

Find the area of each triangle.

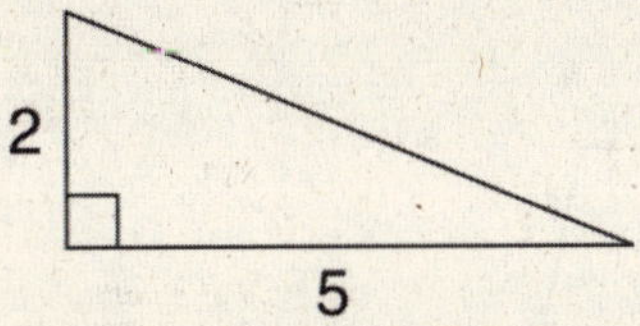

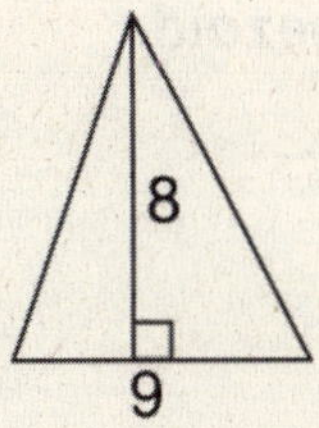

1. $A = \frac{1}{2} \cdot b \cdot h$

   $A = \frac{1}{2} \cdot$ ____ $\cdot$ ____

   $A = \frac{1}{2} \cdot$ ____

   $A =$ ____

   The area of the triangle is ____ units^2.

2. $A = \frac{1}{2} \cdot b \cdot h$

   $A = \frac{1}{2} \cdot$ ____ $\cdot$ ____

   $A = \frac{1}{2} \cdot$ ____

   $A =$ ____

   The area of the triangle is ____ units^2.

3.

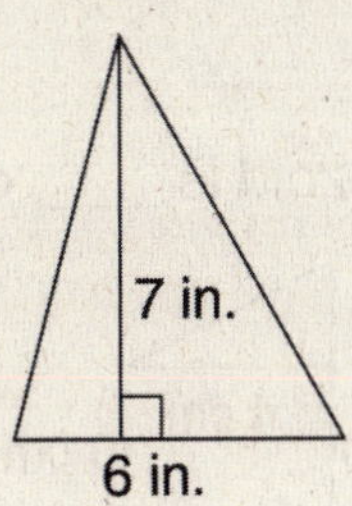

____

4.

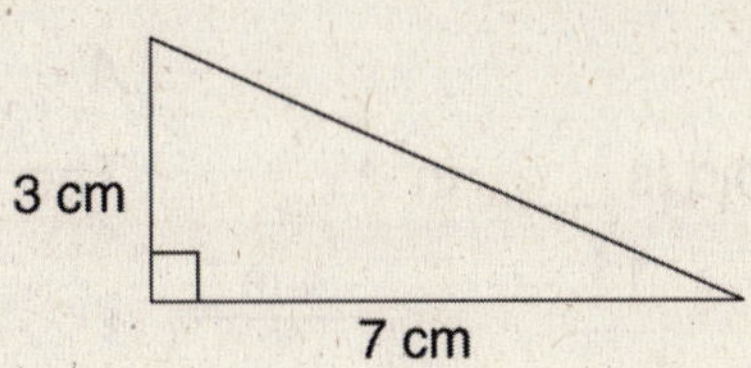

____

5.

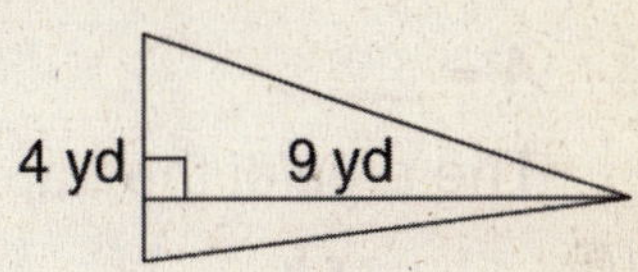

____

6. What is the area of a triangle with base 16 m and height 10 m? ____________

7. What is the area of a triangle with base 25 mm and height 50 mm? ____________

Name ______________________ Date ____________ Class ____________

LESSON 9-4

Review for Mastery

Area of Triangles and Trapezoids (continued)

In a trapezoid, the parallel sides are called the *bases*. One base is always longer than the other. The bases are labeled base 1 and base 2.

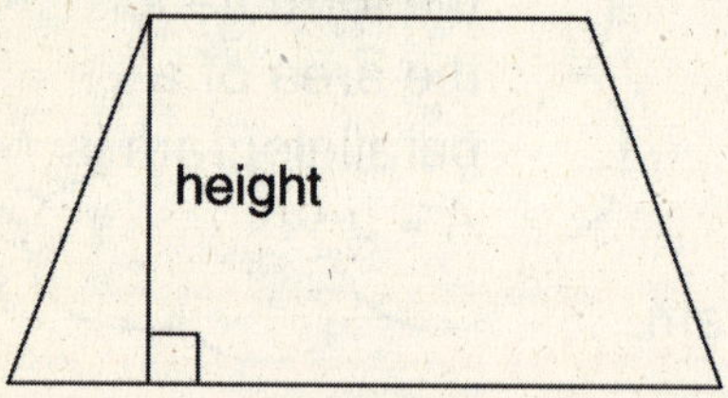

Area of trapezoid = $\frac{1}{2}h(b_1 + b_2)$

Find the area of each trapezoid.

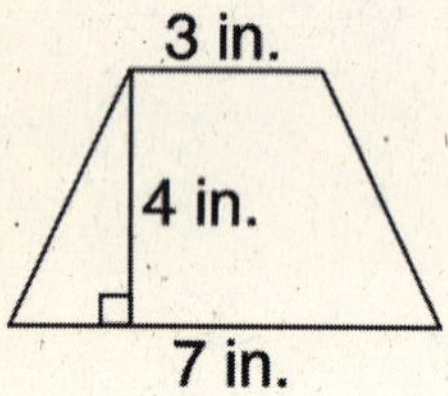

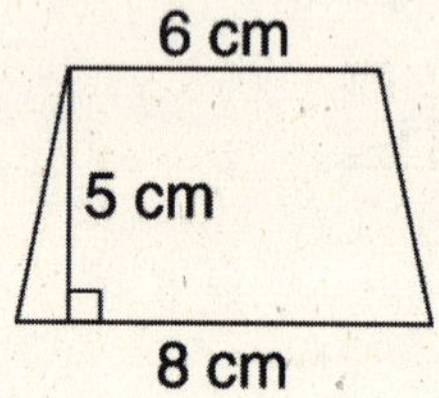

8. $A = \frac{1}{2}h(b_1 + b_2)$

$A = \frac{1}{2} \cdot$ ____ (____ + ____)

$A = \frac{1}{2} \cdot$ ____ (____)

$A = \frac{1}{2} \cdot$ ____

$A =$ ____

The area of the trapezoid is ____ in^2.

9. $A = \frac{1}{2}h(b_1 + b_2)$

$A = \frac{1}{2} \cdot$ ____ (____ + ____)

$A = \frac{1}{2} \cdot$ ____ (____)

$A = \frac{1}{2} \cdot$ ____

$A =$ ____

The area of the trapezoid is ____ cm^2.

10.

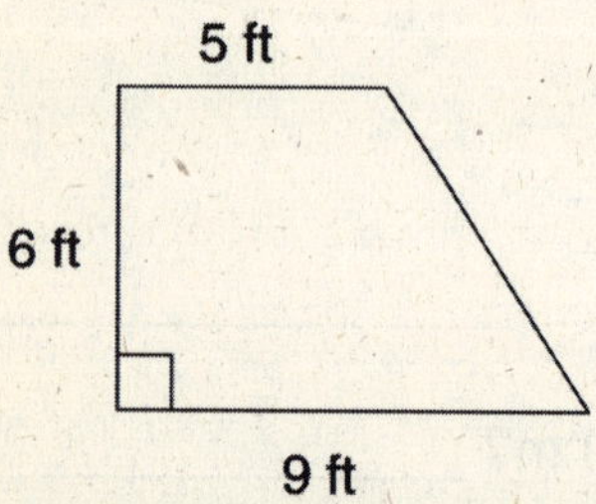

____

11.

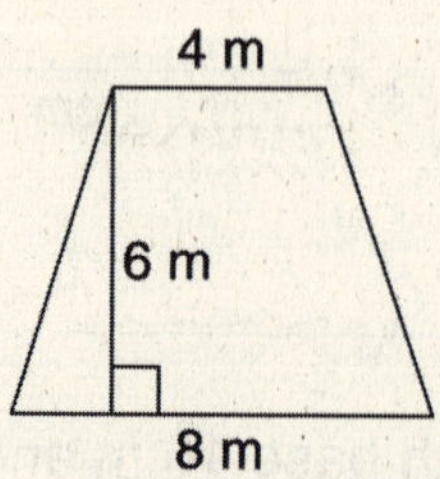

____

12.

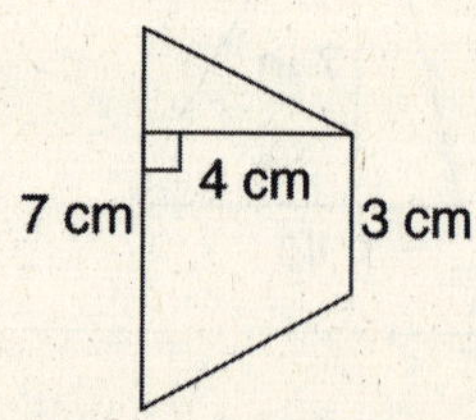

____

13. What is the area of a trapezoid with bases 25 yd and 75 yd and height 10 yd?

____

Name ______________________ Date ______________ Class ______________

LESSON 9-4

Challenge

Break It Up

One way to find the area of this figure is to divide it into a rectangle and a triangle. Find the area of each, then add the areas together.

Area of rectangle = $7 \cdot 10 = 70 \text{ m}^2$

Area of triangle = $\frac{1}{2}(5 \cdot 10) = 25 \text{ m}^2$

Total area of figure = Area of rectangle + Area of triangle

$= 70 \text{ m}^2 + 25 \text{ m}^2 = 95 \text{ m}^2$

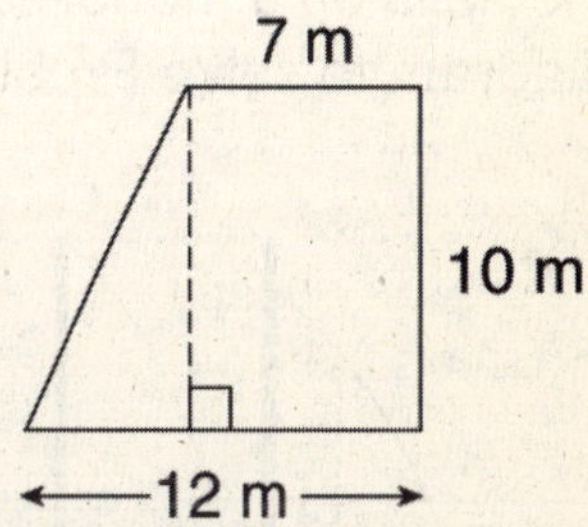

Divide each figure into parts. Then find the area of each part. Add the areas of the parts to find the area of the whole figure.

1.

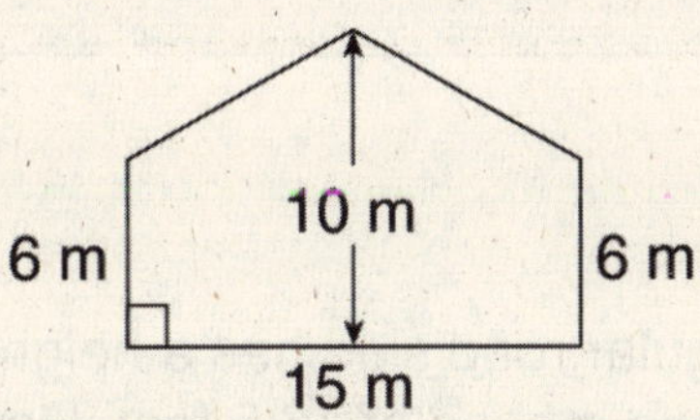

____

2.

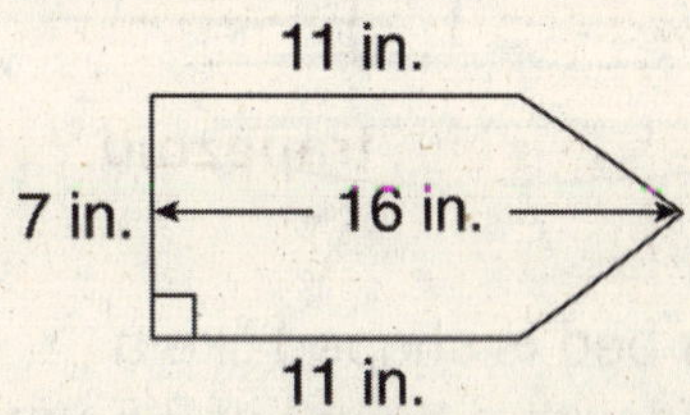

____

3.

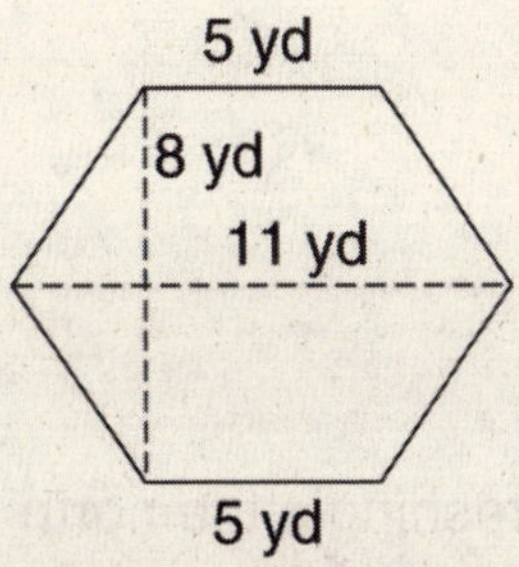

____

4.

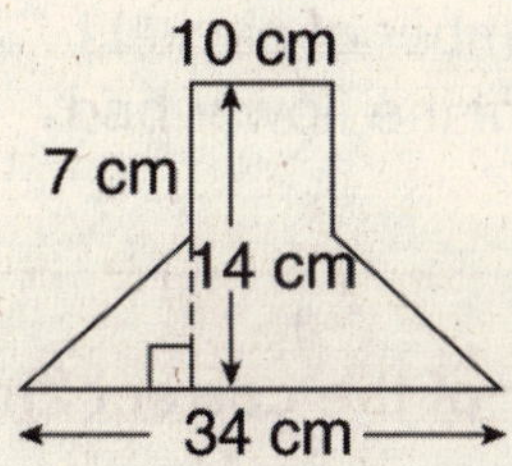

____

For Exercises 5 and 6, you need to *subtract* the area of a part to find the area of each whole figure.

5.

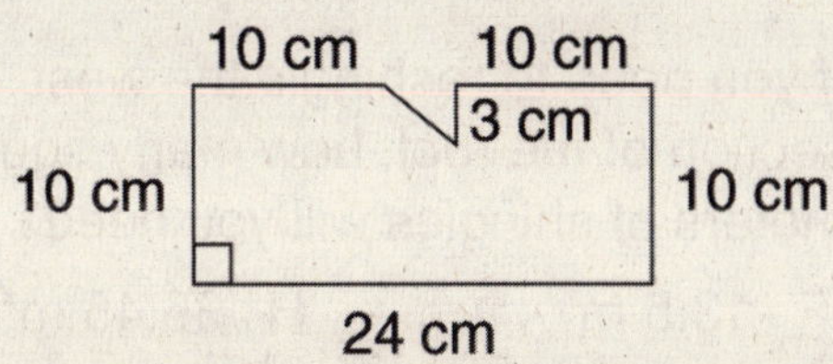

____

6.

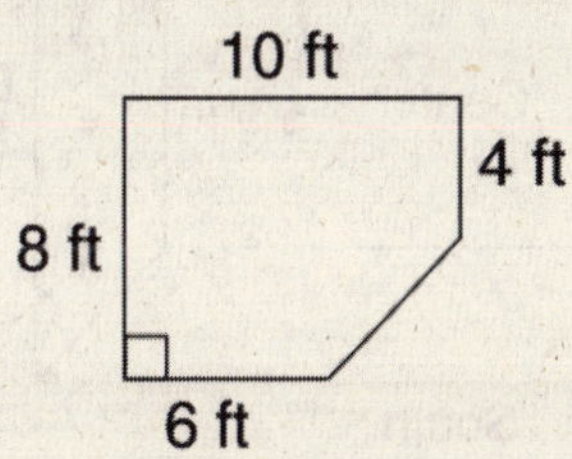

____

Name ______________________ Date ______________ Class ______________

LESSON 9-4

Problem Solving

Area of Triangles and Trapezoids

Write the correct answer.

The diagram shows the dimensions of the sails on a model sailboat. Use the diagram to solve Problems 1–2.

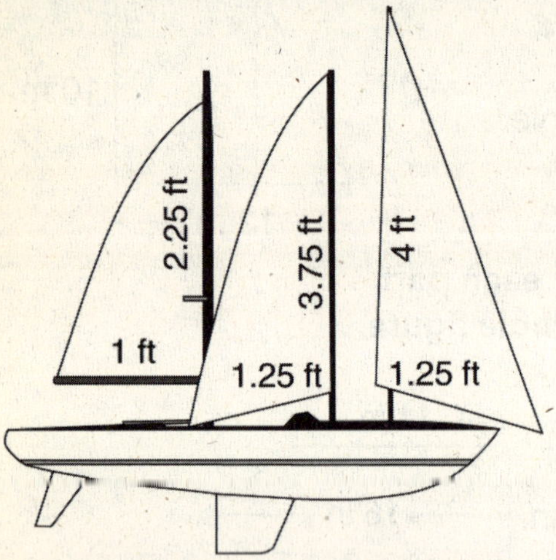

1. About how much material to the nearest square foot will be needed to make the sails?

____

2. If the dimensions for each sail were doubled, how would that change the total amount of material needed to make the sails?

____

____

3. A flower bed is shaped like a trapezoid with a height of 3.5 yards, one 2.8-yard base, and another 4.6-yard base. A packet of flower seeds covers 5.6 square yards. What is the least number of packets needed to plant the flower bed?

____

4. A triangular road sign has a height of 8 feet and a base of 16.5 feet. How much larger in area is this sign than one with a height of 4 feet and a base of 8.25 feet?

____

Choose the letter of the correct answer.

This diagram shows the top view of the roof of a house.

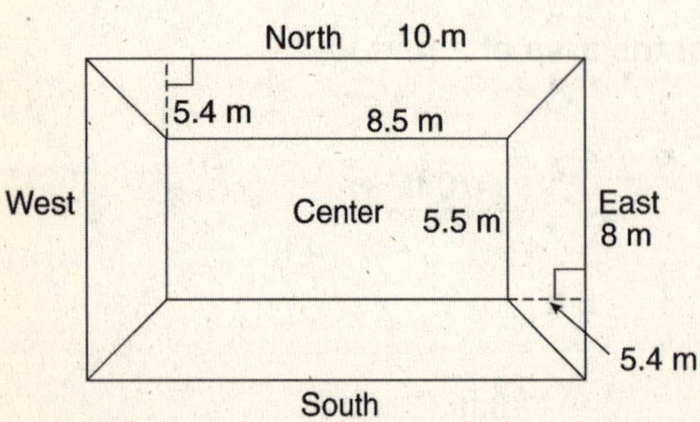

5. If you need to reshingle the north and south sections of the roof, how many square meters of shingles will you need?

A 199.8 m^2 C 49.95 m^2

B 99.9 m^2 D 459 m^2

6. If you need to reshingle the west section of the roof, how many square meters of shingles will you need?

F 13.5 m^2 H 36.45 m^2

G 18.9 m^2 J 72.9 m^2

Name ______________________ Date ______________ Class ______________

LESSON 9-4

Reading Strategies

Use Graphic Aid

Knowing how to find the area of a square or parallelogram can help you find the area of a triangle.

When you draw a diagonal line through a square, you make two equal triangles.

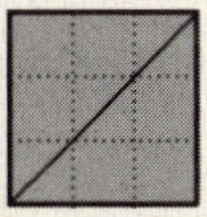

Answer each question.

1. How many square units are in the square? ______________

2. To find out how many squares are in one of the triangles, count the number of half squares first. How many did you find? ______________

3. How many whole squares can you count? ______________

4. How many half and whole squares together? ______________

You can also draw a diagonal line through a parallelogram and divide it into two equal triangles.

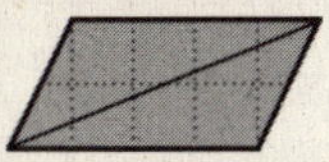

Use the figure to answer each.

5. How many square units are in the parallelogram? ______________

6. Count the full squares and half squares in one triangle and write the total. ______________

7. Compare the area of a parallelogram or square with the area of a triangle.

____

Name ______________________ Date ______________ Class ______________

LESSON 9-4

Puzzles, Twisters & Teasers

All Washed Up!

Find the area of each figure below. Match the letters to solve the riddle.

1. ______________ **S**

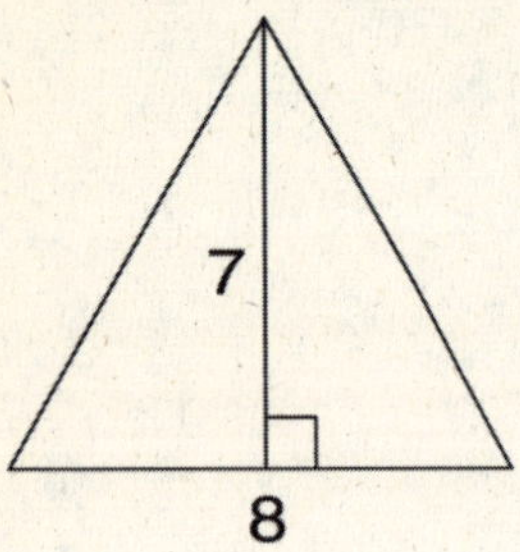

2. ______________ **M**

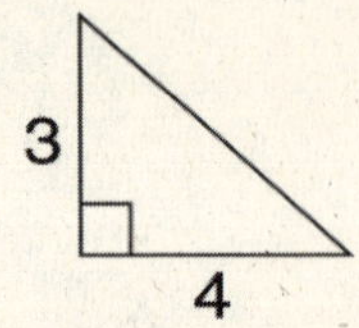

3. ______________ **E**

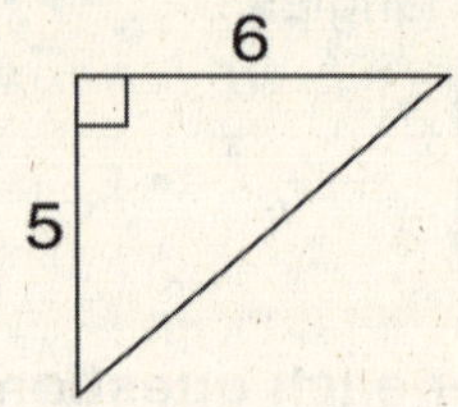

4. ______________ **I**

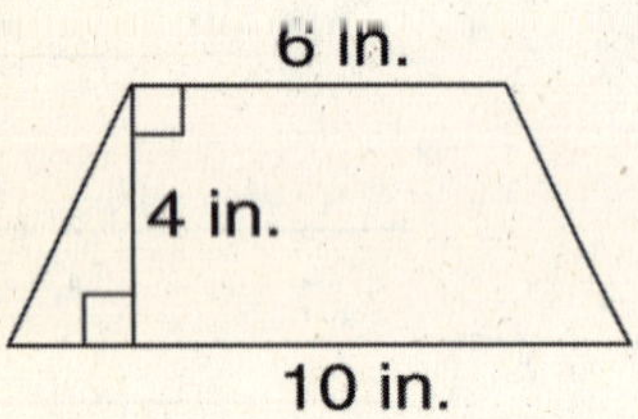

5. ______________ **V**

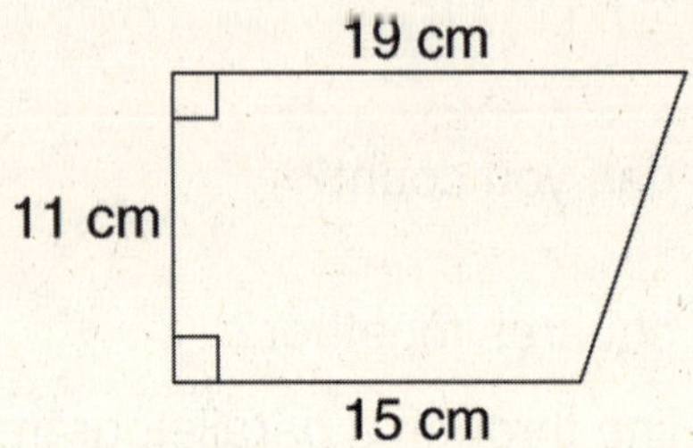

6. ______________ **C**

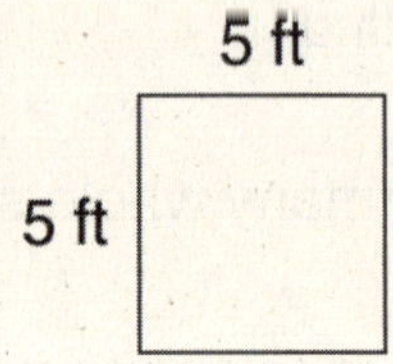

7. ______________ **A**

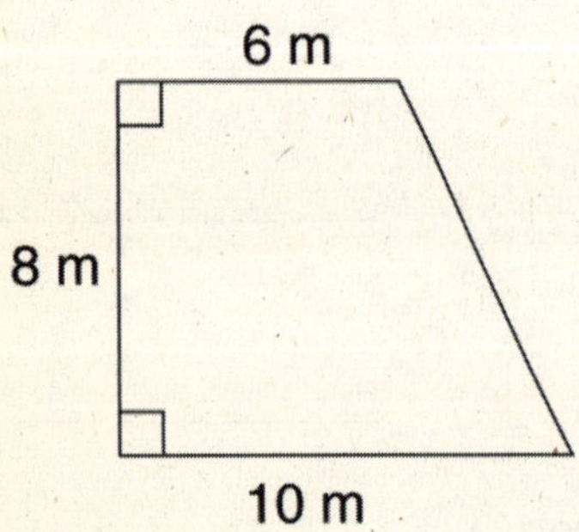

8. ______________ **R**

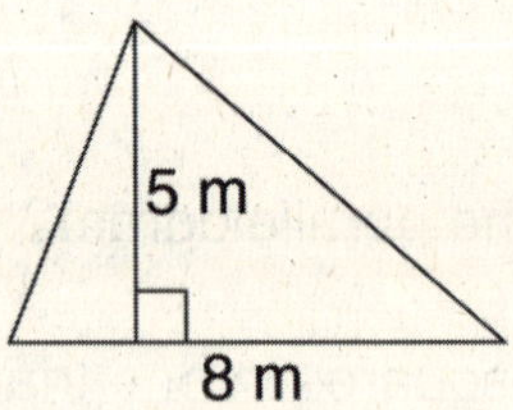

9. ______________ **W**

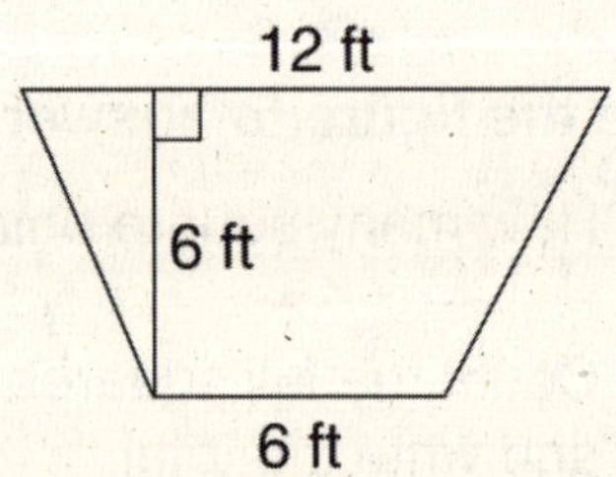

What washes up on small beaches?

___	___	___	___	**O**	-	___	___	___	___	___
6	32	25	20			54	64	187	15	28

Name ______________________ Date ______________ Class ______________

LESSON 9-5

Practice A

Area of Circles

Find the area of each circle to the nearest tenth. Use 3.14 for π. Cross out each number in the box that matches an area.

19.6	254.3	50.2	78.5	22.0	260.0	
153.9	28.3	3.1	379.9	171.9	72.3	176.6

1.
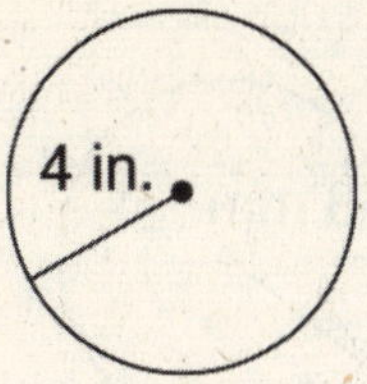

____

2.
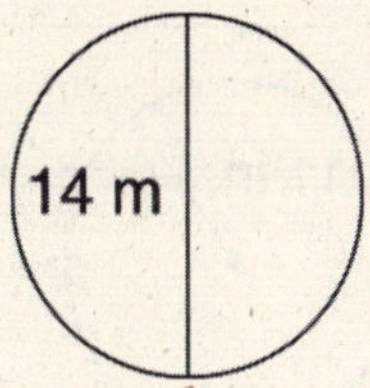

____

3.
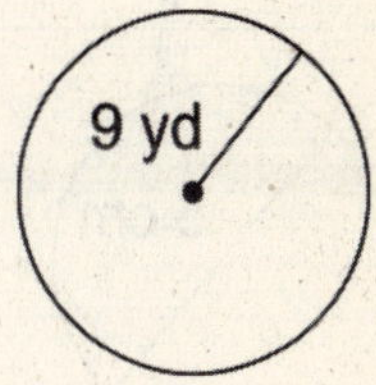

____

4.
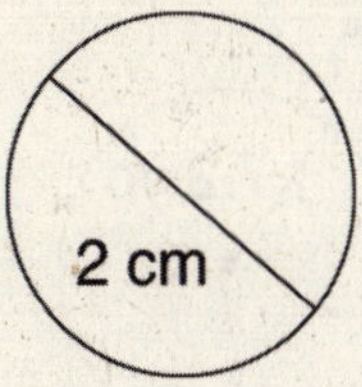

____

5.
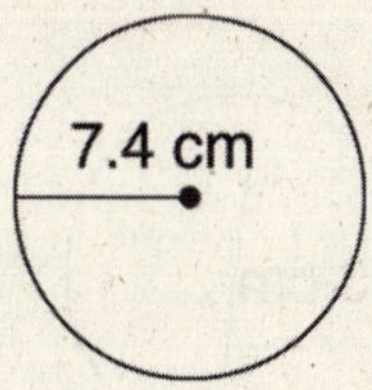

____

6.
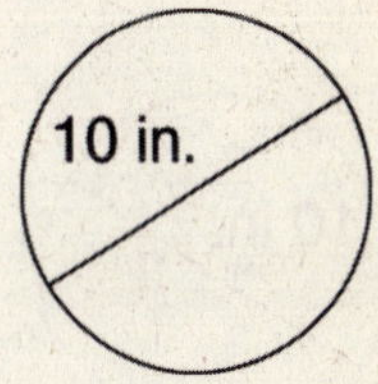

____

7.
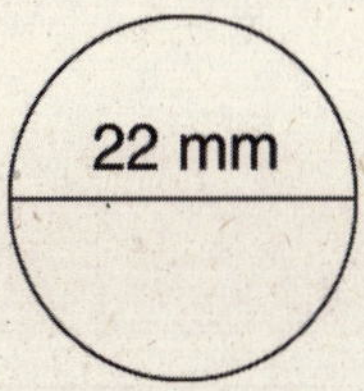

____

8.
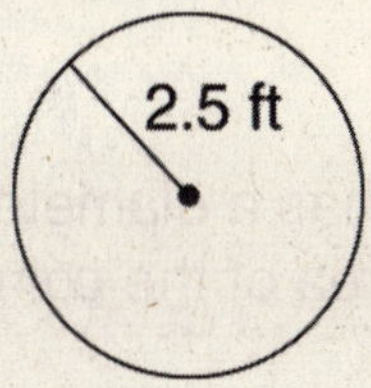

____

9.
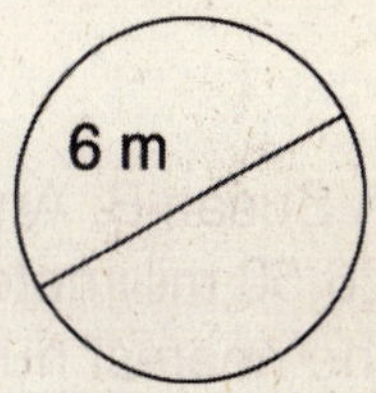

____

10. A round table has a diameter of 28 inches. What is the area of the tabletop? Use $\frac{22}{7}$ for π.

____

11. Find the area of the shaded region of the circle. Use 3.14 for π. Round your answer to the nearest tenth.

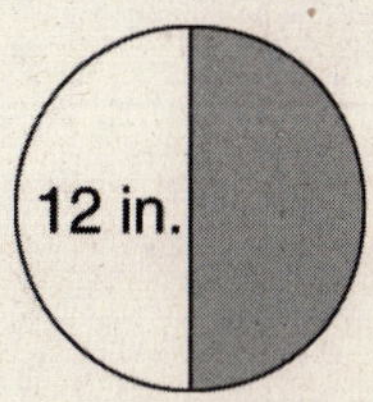

____

Name ______________________ Date ______________ Class ______________

LESSON 9-5

Practice B

Area of Circles

Find the area of each circle to the nearest tenth. Use 3.14 for π.

1.
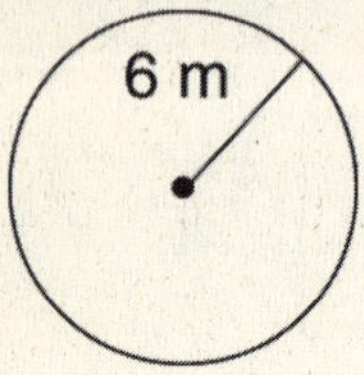

____

2.
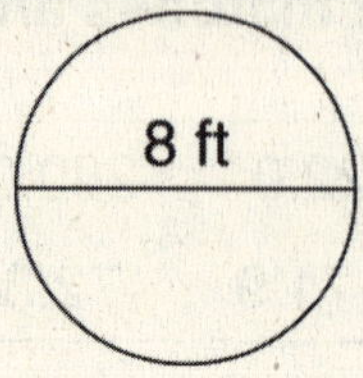

____

3.
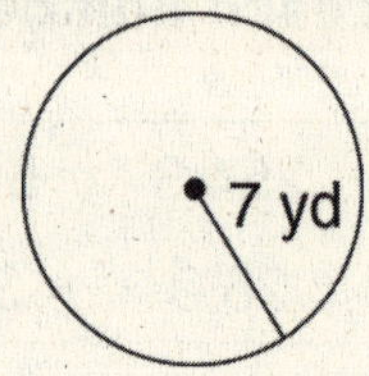

____

4.
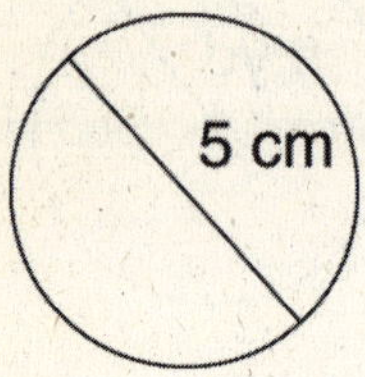

____

5.

____

6.
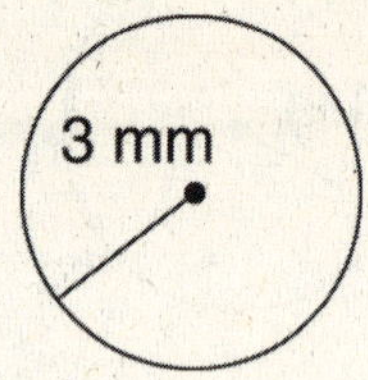

____

7.
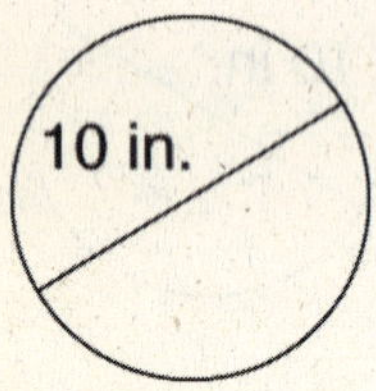

____

8.
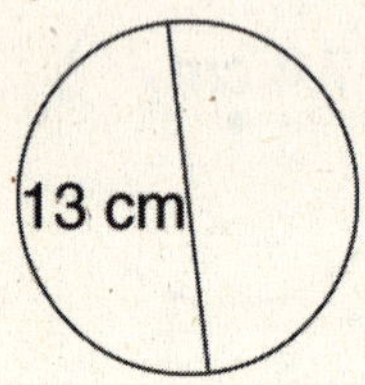

____

9.
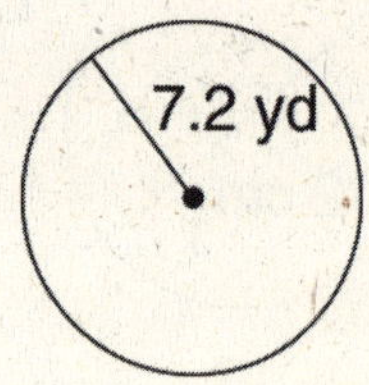

____

10. A Susan B. Anthony dollar coin has a diameter of 26.50 millimeters. What is the area of the coin to the nearest hundredth?

____

11. A tablecloth for a round table has a radius of 21 inches. What is the area of the tablecloth? Use $\frac{22}{7}$ for π.

____

12. Use a centimeter ruler to measure the radius of the circle. Then find the area of the shaded region of the circle. Use 3.14 for π. Round your answer to the nearest tenth.

____

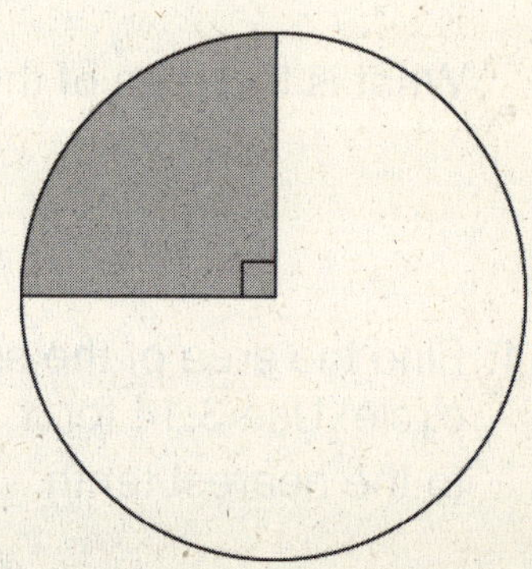

Name ______________________ Date ______________ Class ______________

LESSON 9-5

Practice C

Area of Circles

Find the area of each circle to the nearest tenth. Use 3.14 for π.

1. 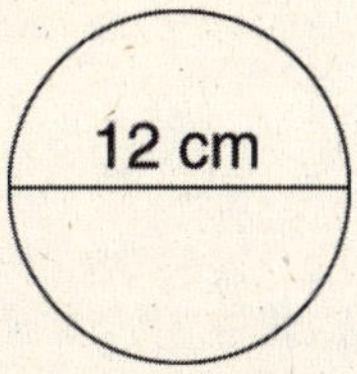

____

2. 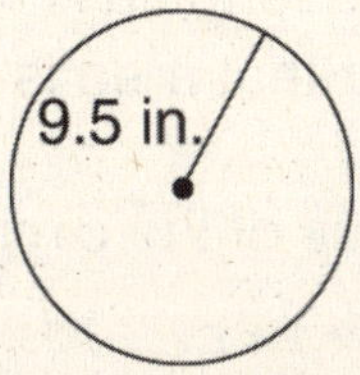

____

3. 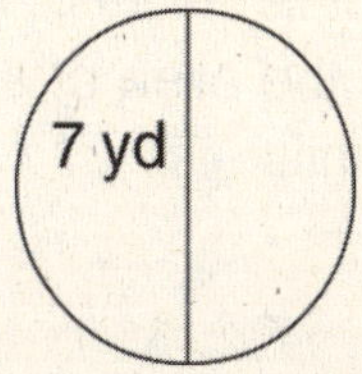

____

4. 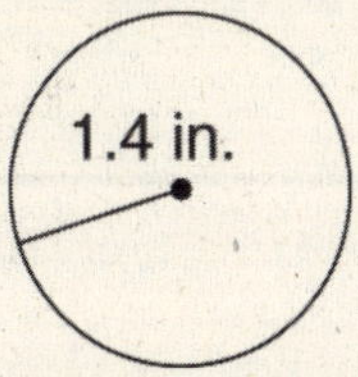

____

5. 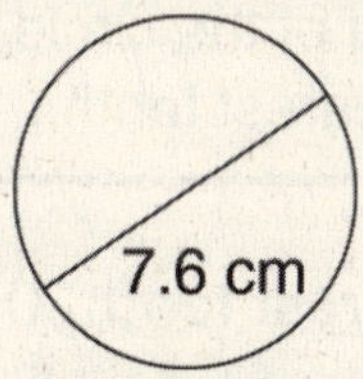

____

6. 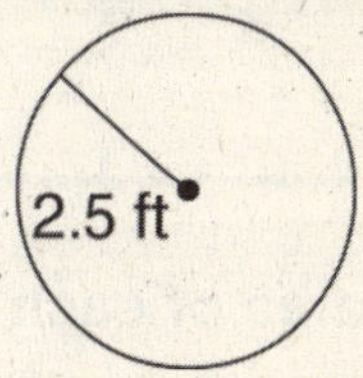

____

Given the radius or diameter, find the circumference and area of each circle to the nearest tenth. Use 3.14 for π.

7. $r = 4$ in.

____

____

8. $d = 14$ cm

____

____

9. $d = 26$ ft

____

____

Given the area, find the radius of each circle. Use 3.14 for π.

10. $A = 200.96\ \text{m}^2$

____

11. $A = 12.56\ \text{in}^2$

____

12. $A = 314\ \text{yd}^2$

____

13. A playground area is circular with a diameter of 32 feet. What is the area of the playground? Round your answer to the nearest tenth.

____

14. Use a centimeter ruler to measure the radius of the circle. Then find the area of the shaded region of the circle. Use 3.14 for π. Round your answer to the nearest tenth.

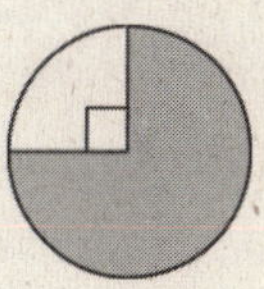

____

Name ____________________ Date ____________ Class ____________

LESSON 9-5

Review for Mastery

Area of Circles

The formula $A = \pi r^2$ is used to find the area of a circle. Since the value of π is about 3.14, you can use the formula $A \approx 3.14 \cdot r \cdot r$ to estimate the area of a circle. Remember that area is expressed in square units.

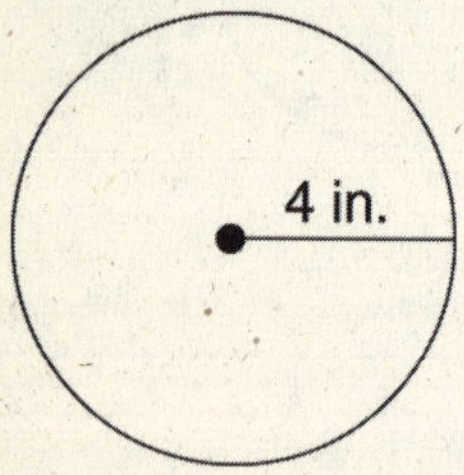

The radius of the circle is 4 in.

$A \approx 3.14 \cdot r \cdot r$

$A \approx 3.14 \cdot 4 \cdot 4$

$A \approx 50.24$

The area of the circle is 50.2 in^2 to the nearest tenth.

Find the area of each circle to the nearest tenth. Use 3.14 for π.

1.

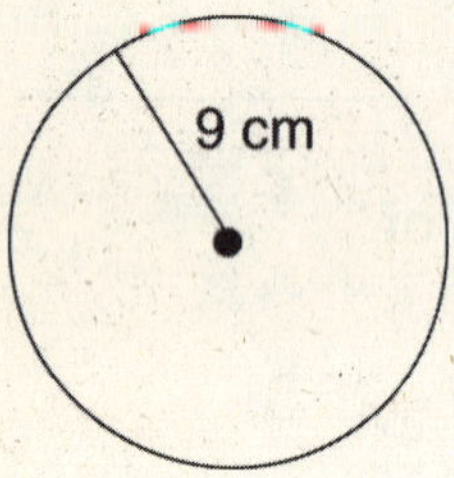

The radius is ______ cm.

$A = \pi r^2$

$A \approx 3.14 \cdot$ ______ $\cdot$ ______

$A \approx$ ________

The area is ________ cm^2 to the nearest tenth.

2.

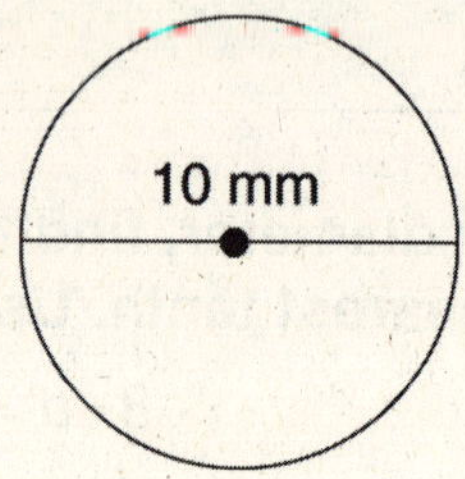

The diameter is 10 mm.

The radius is ______ mm.

$A = \pi r^2$

$A \approx 3.14 \cdot$ ______ $\cdot$ ______

$A \approx$ ________

The area is ________ mm^2 to the nearest tenth.

3.

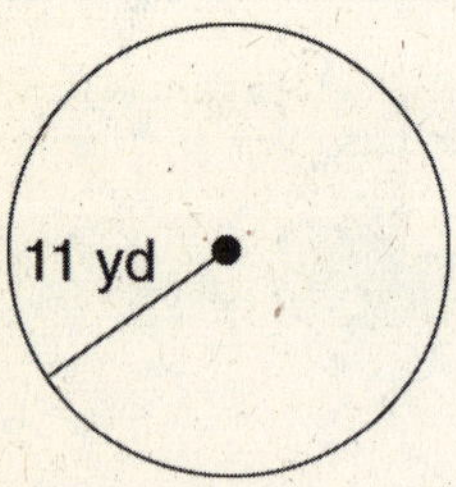

____

4.

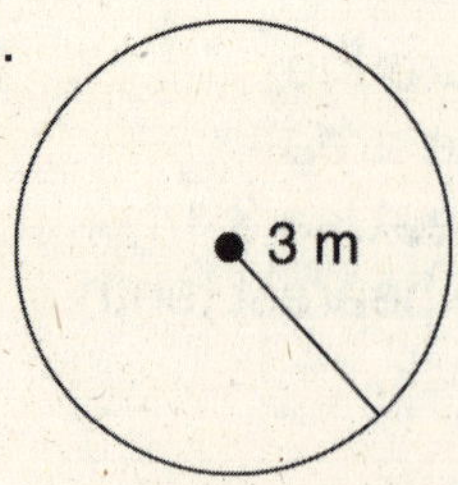

____

5.

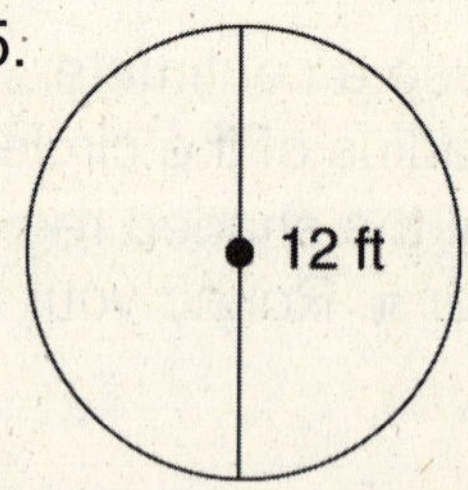

____

6. What is the area of a circle with radius 13 yd? Round your answer to the nearest tenth.

____

Name ______________________ Date ______________ Class ______________

LESSON 9-5

Challenge

Part of the Picture

You can add or subtract to find the area of a shaded region.

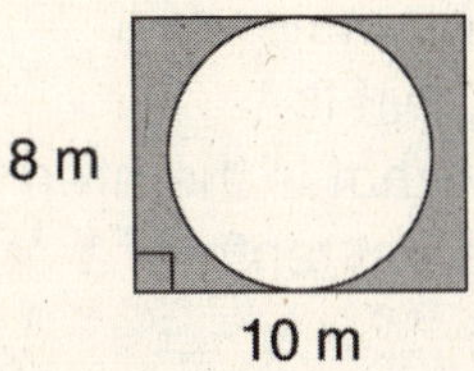

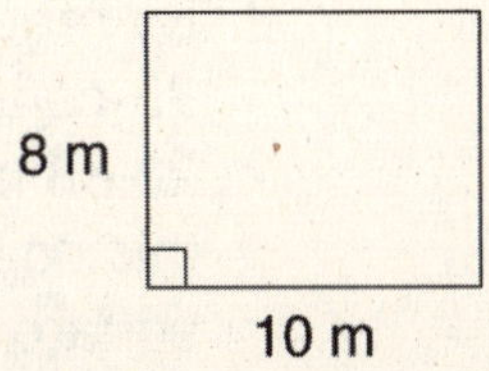

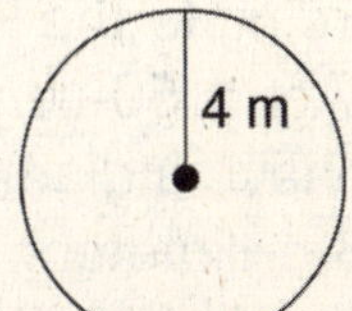

Find the area of the rectangle.
$A = \ell\omega = 10 \cdot 8 = 80 \text{ m}^2$

Find the area of the circle.
$A = \pi r^2 = \pi \cdot 4^2 \approx 50.24 \text{ m}^2$

The shaded area $\approx 80 \text{ m}^2 - 50.24 \text{ m}^2 \approx 29.76 \text{ m}^2 \approx 29.8 \text{ m}^2$.

Add or subtract to find the area of the shaded region. Round your answer to the nearest tenth.

1. 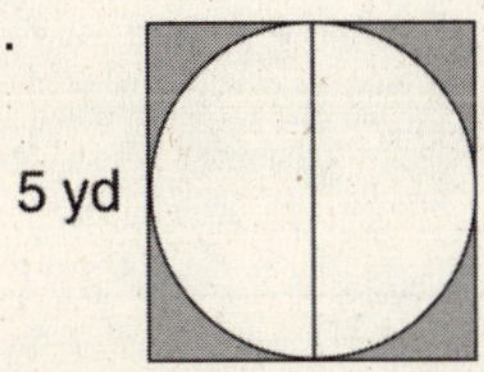

____

2. 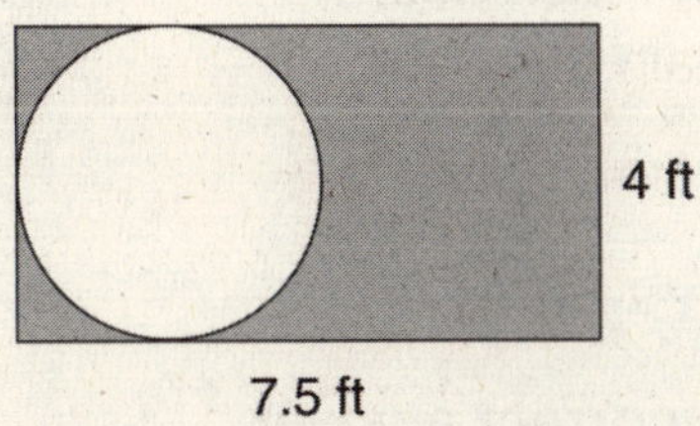

____

3. 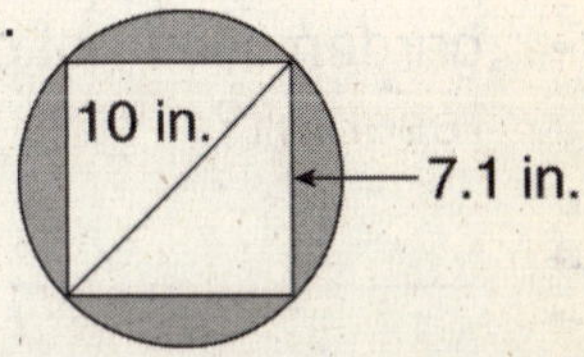

____

4. 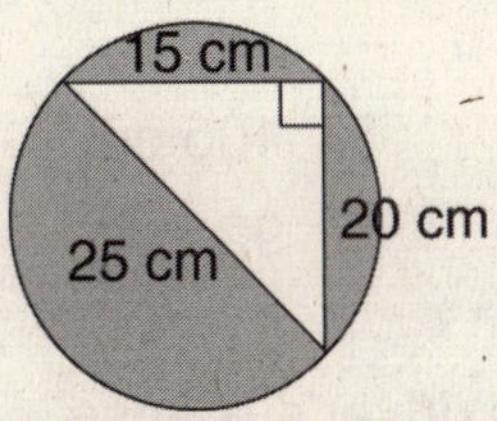

____

5. 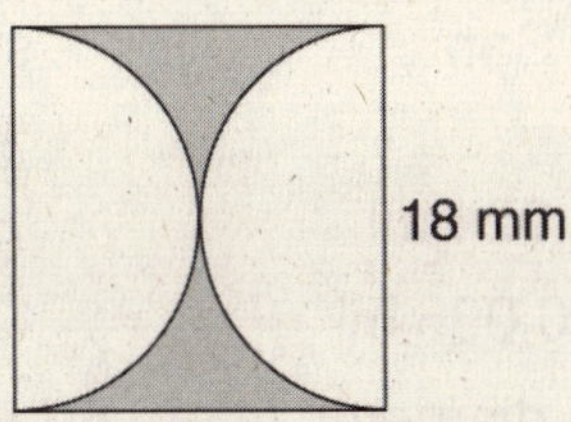

____

6. 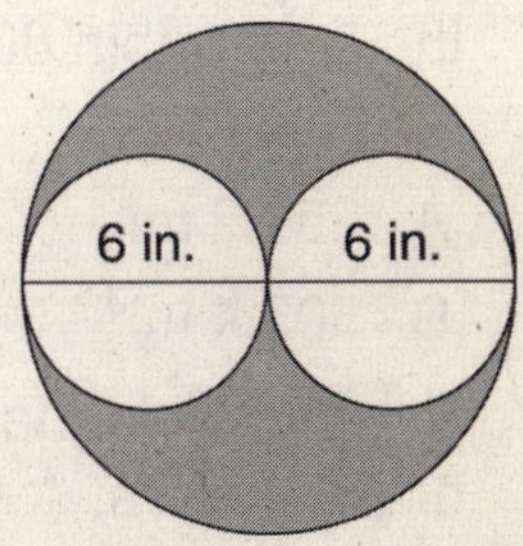

____

7. 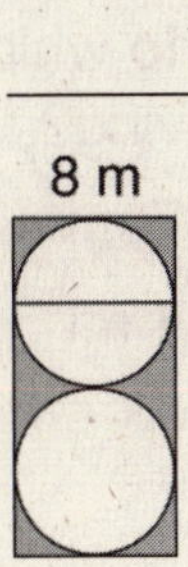

____

8. 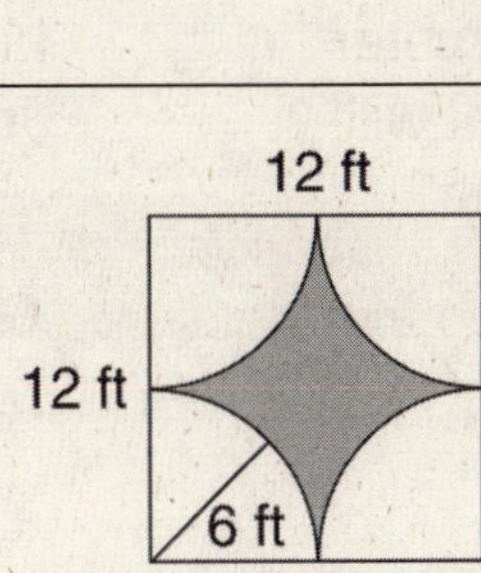

____

9. 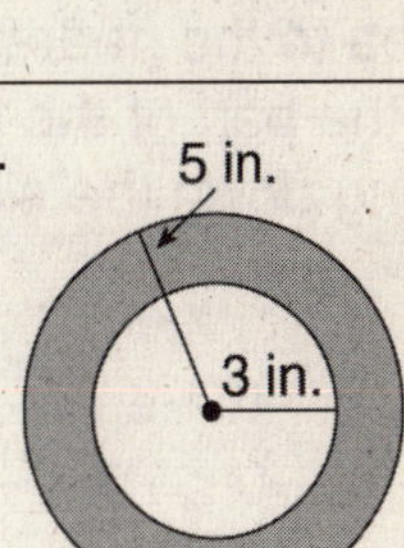

____

Name ______________________ Date ______________ Class ______________

LESSON 9-5

Problem Solving

Area of Circles

Write the correct answer.

1. According to the Royal Canadian Mint Act, a 50-cent Canadian coin must have a diameter of 27.13 millimeters. What is the area of this coin to the nearest tenth of a square millimeter?

____

2. By regulation, the diameter of a 25-cent Canadian coin is 23.88 millimeters. What is the area of this coin to the nearest tenth of a square millimeter?

____

3. There is a water reservoir beneath a circular garden to supply a fountain in the garden. The reservoir has a 26-inch diameter. The garden has a 12-foot diameter. How much of the garden does not contain the water reservoir?

____

4. A frying pan has a diameter of 11 inches. What is the area to the nearest square inch of the smallest cover that will fit on top of the frying pan?

____

Choose the letter of the correct answer.

5. In the state of Texas, Austin is about 80 miles northeast of San Antonio. What area is represented by all of the land within 80 miles of San Antonio?

   A 251.2 mi^2 C 5,024 mi^2

   B 502.4 mi^2 D 20,096 mi^2

6. A standard CD has a diameter of 12 centimeters. What is the area of a circular case that can be used to store a CD?

   F 114 cm^2 H 105 cm^2

   G 112 cm^2 J 92 cm^2

7. A round dining table has a diameter of 2.5 meters. A round tablecloth has a diameter of 3.5 meters. What is the area to the nearest tenth of a meter of the part of the tablecloth that will hang down the side of the table?

   A 18.8 m^2 C 4.7 m^2

   B 6.3 m^2 D 1.0 m^2

8. Justin just got his driver's license. His parents are giving him permission to drive within a 25-mile radius of his home. What is the area to which Justin is restricted when driving?

   F 7,850 mi^2 H 157 mi^2

   G 1,962.5 mi^2 J 314 mi^2

Name ______________________ Date ______________ Class ______________

LESSON 9-5

Reading Strategies

Use Graphic Aid

You can use what you know about the area of parallelograms to understand the area of circles.

The formula for the area of a parallelogram is:
Area = bh

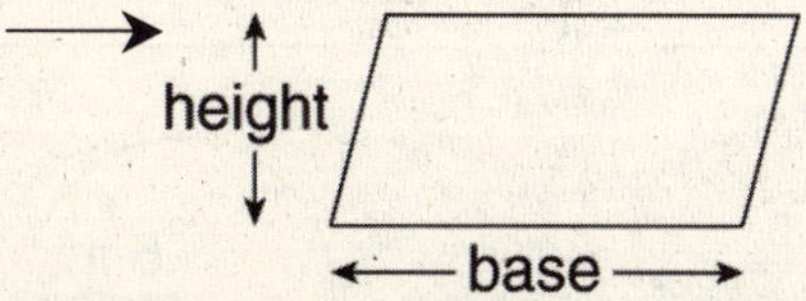

You can cut a circle into wedges and place the wedges next to each other to approximate a parallelogram.

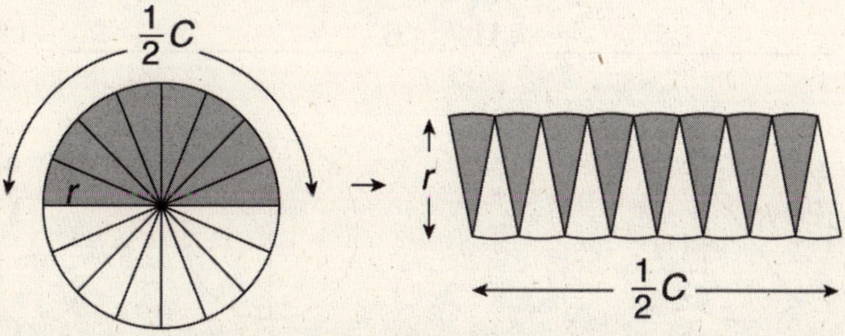

Compare the parallelogram to the circle to answer each question.

1. What figure do the wedges of a circle approximately form?

____

2. What is the formula for finding the area of a parallelogram?

____

3. A parallelogram has a measurement called the *base*. When a circle is reshaped into a parallelogram, what measurement is the same as the base?

____

4. The other measurement of a parallelogram is the height. When a circle is reshaped into a parallelogram, what measurement is the same as the height?

____

5. Use the information from Exercises 3 and 4 and the rearranged parallelogram to write the formula for the area of a circle.

____

Name ____________________ Date ____________ Class ____________

LESSON 9-5

Puzzles, Twisters & Teasers

Area of Circles

Find the area of each circle and write it on the line. Round to the nearest hundredth. Use 3.14 for π. Use the answers to solve the riddle.

1. ________________ = **E**

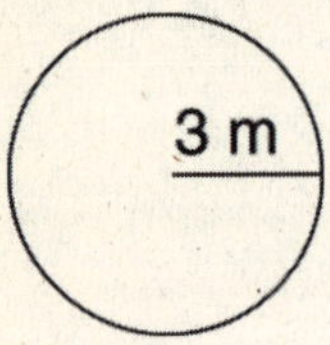

2. ________________ = **T**

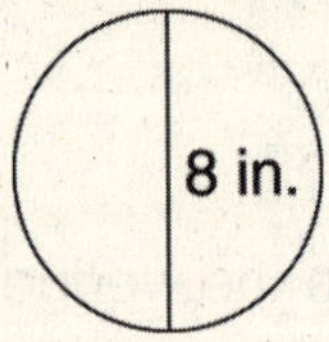

3. ________________ = **S**

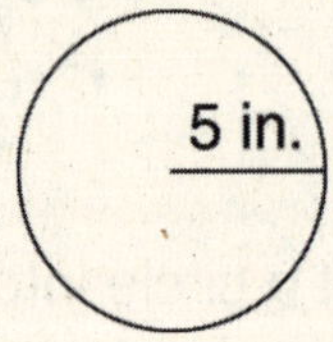

4. ________________ = **L**

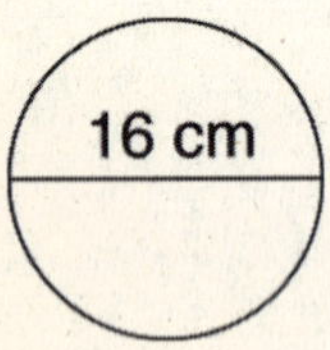

5. ________________ = **U**

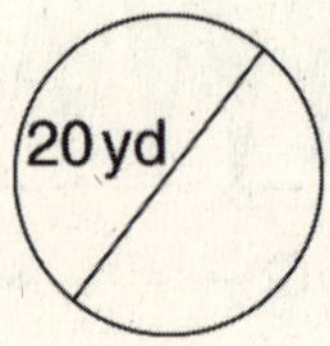

6. ________________ = **I**

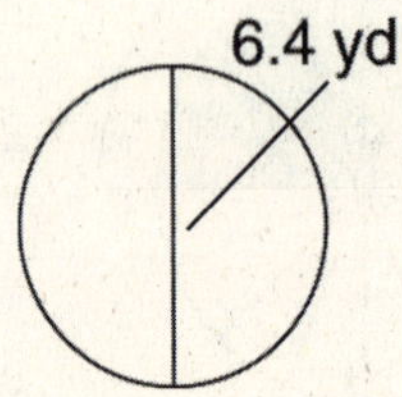

7. ________________ = **O**

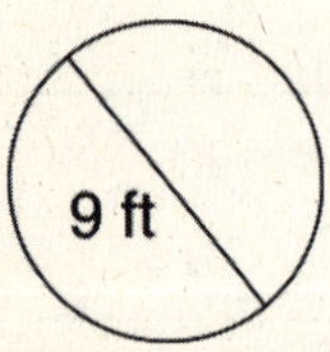

8. ________________ = **G**

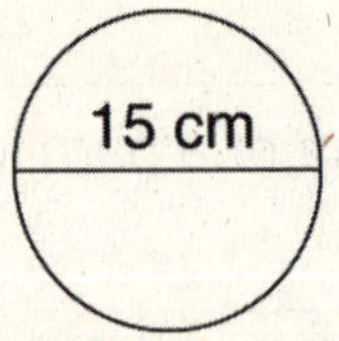

9. ________________ = **H**

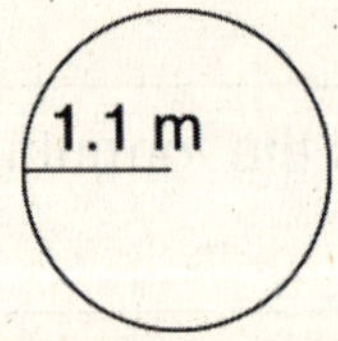

Which kind of house weighs the least?

A ____ ____ ____ ____ ____ ____ ____ ____ ____ ____

200.96 32.15 176.63 3.8 50.24 3.8 63.59 314 78.5 28.26

Name ______________________ Date ______________ Class ______________

LESSON 9-6

Practice A

Area of Irregular Figures

Estimate the area of each figure. Each square represents 1 square foot. Choose the letter for the best answer.

1.

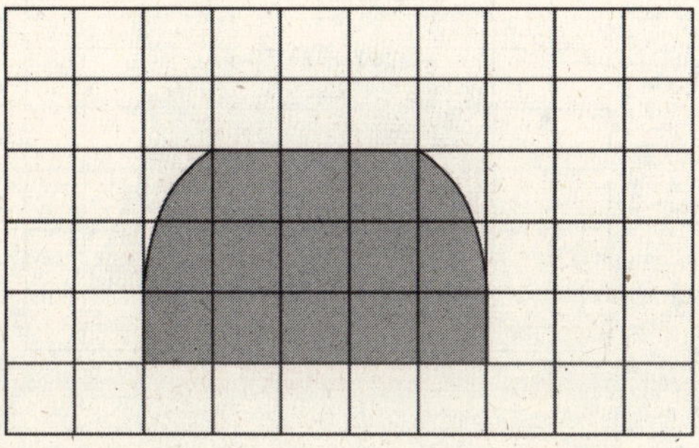

A 10 ft^2　　C 14 ft^2

B 11 ft^2　　D 15 ft^2

2.

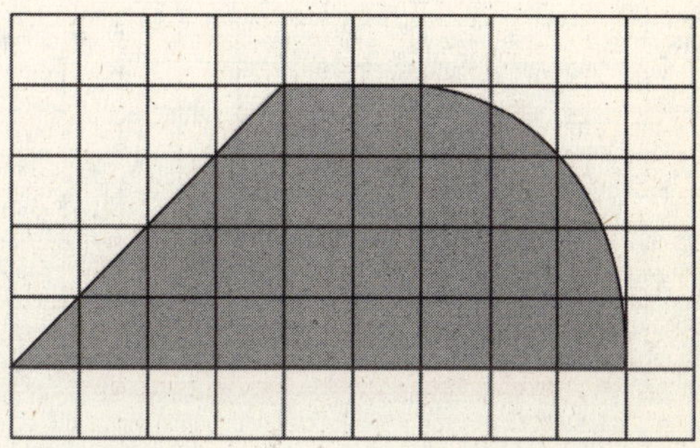

F 24 ft^2　　H 32 ft^2

G 27 ft^2　　J 36 ft^2

Find the area of each figure. Use 3.14 for π.

3.

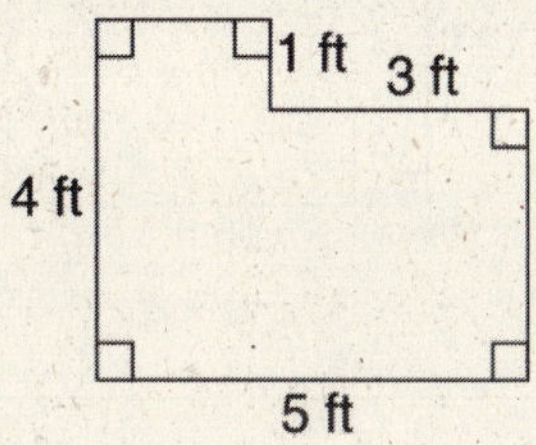

____

4.

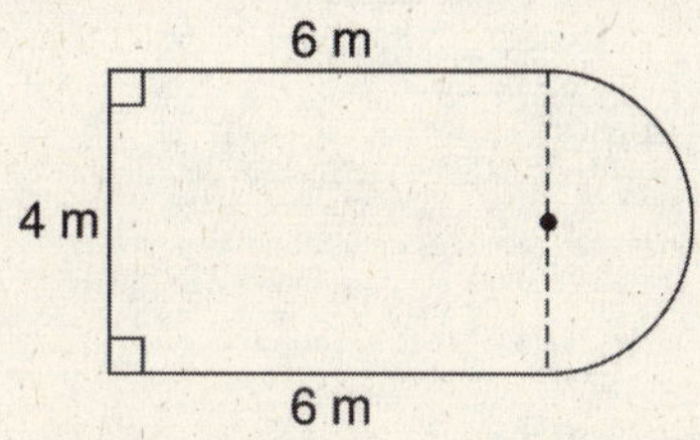

____

5.

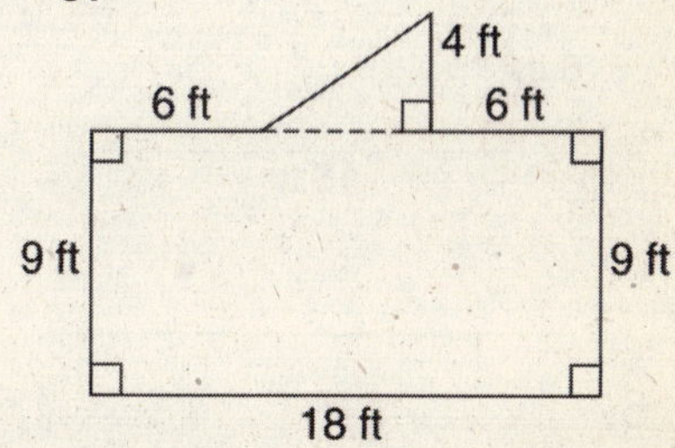

____

6.

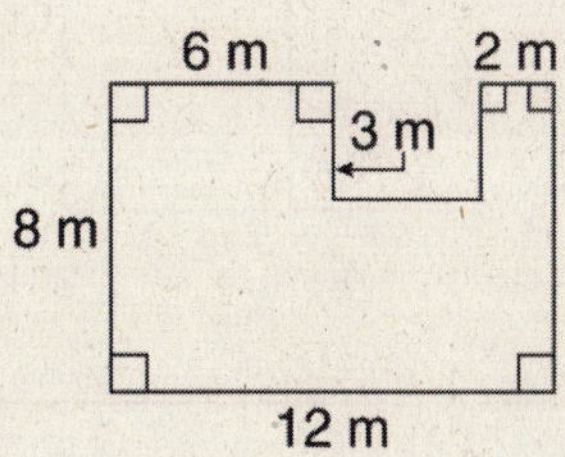

____

7.

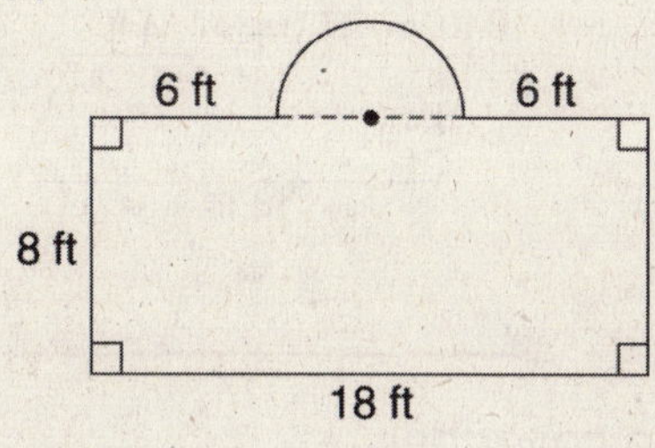

____

8.

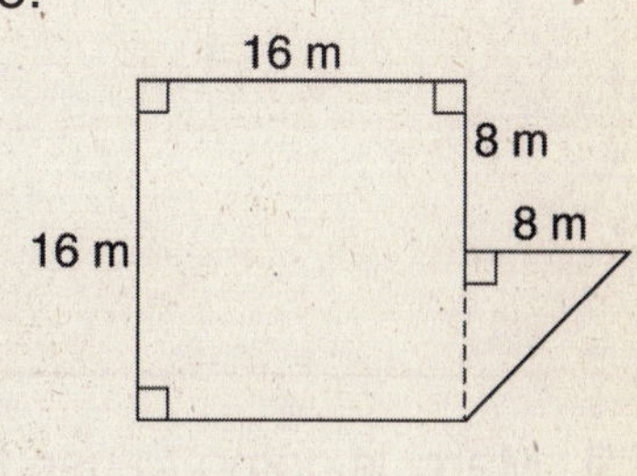

____

9. The figure shows the dimensions of a room. How much carpet is needed to cover the floor?

____

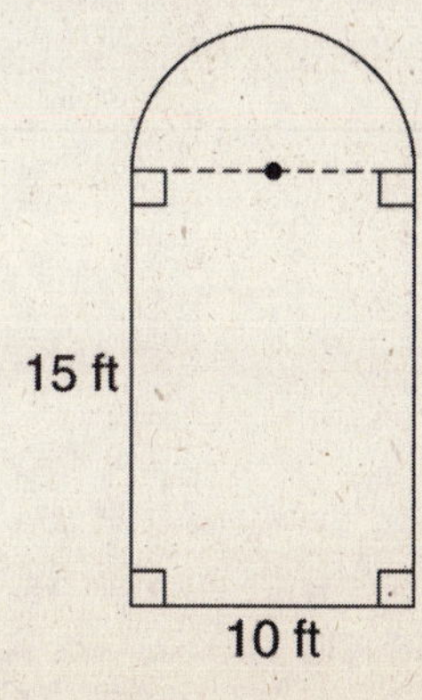

Name ______________________ Date ______________ Class ______________

LESSON 9-6

Practice B

Area of Irregular Figures

Estimate the area of each figure. Each square represents 1 square foot.

1.

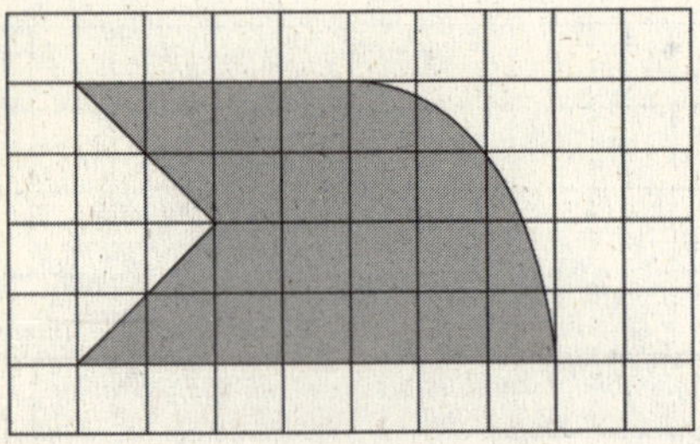

____

2.

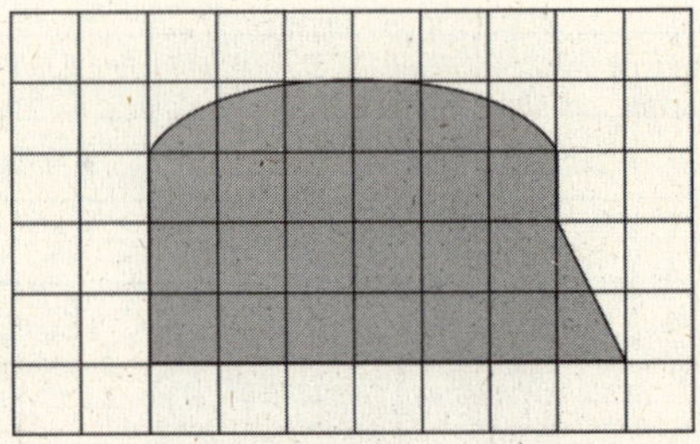

____

Find the area of each figure. Use 3.14 for π.

3.

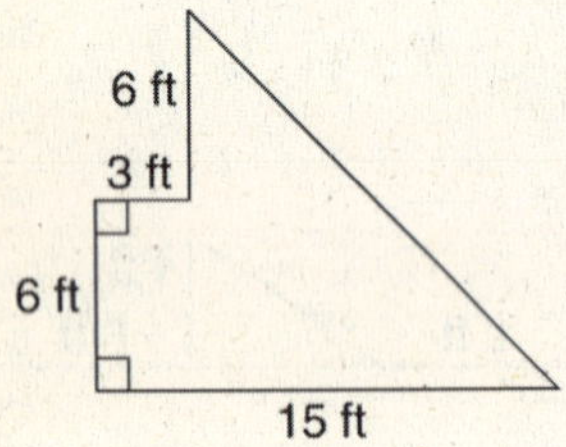

____

4.

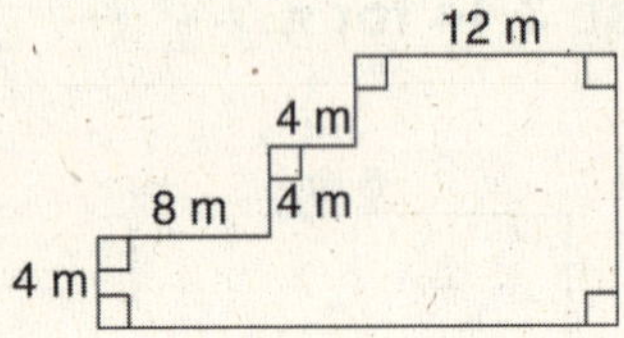

____

5.

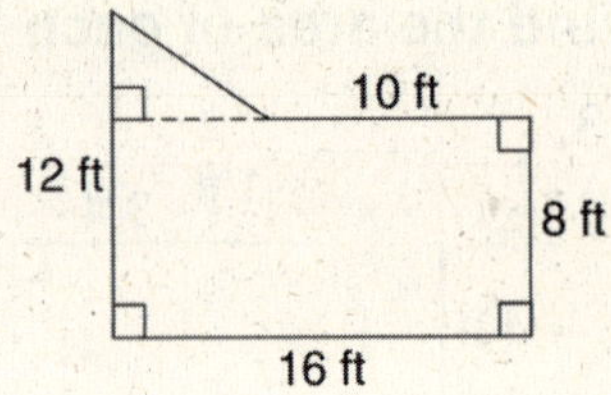

____

6.

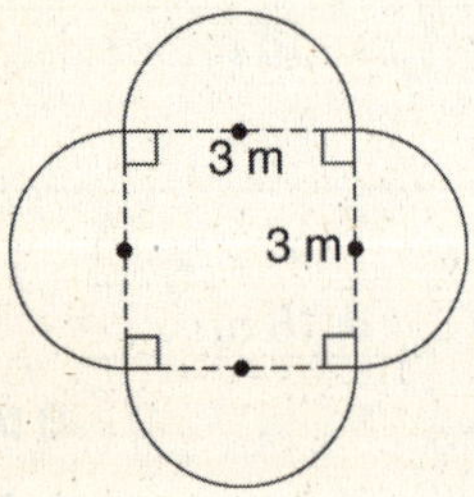

____

7.

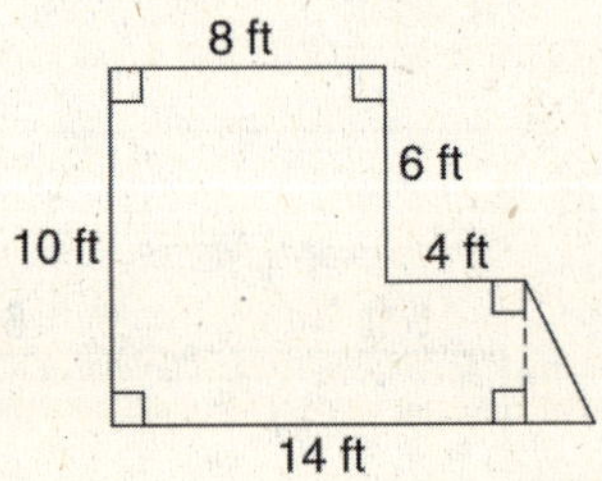

____

8.

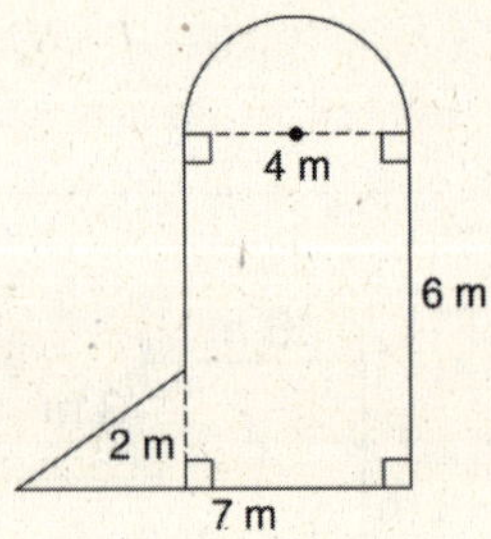

____

9. Marci is going to use tile to cover her terrace. How much tile does she need?

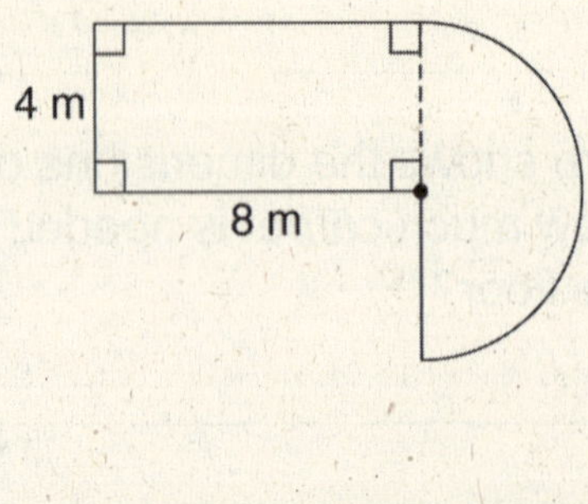

____

Name ______________________ Date ______________ Class ______________

LESSON 9-6

Practice C

Area of Irregular Figures

Estimate the area of each figure. Each square represents 1 square foot.

1. 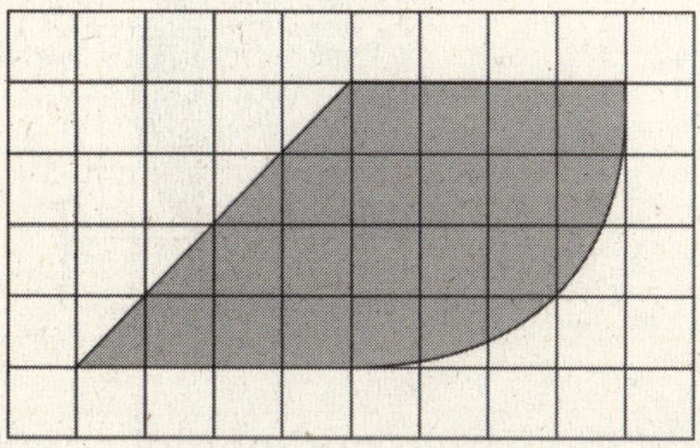

____

2. 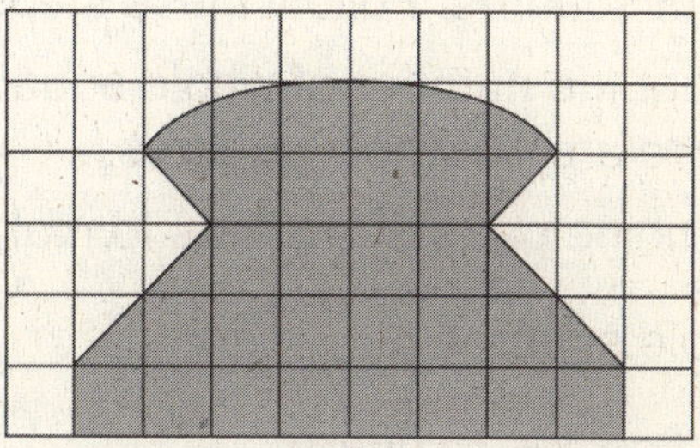

____

Find the area of each figure. Use 3.14 for π.

3. 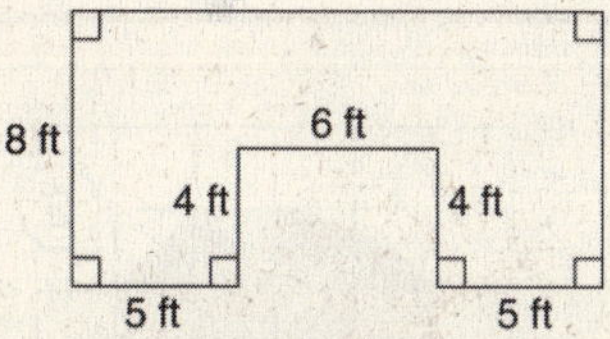

____

4. 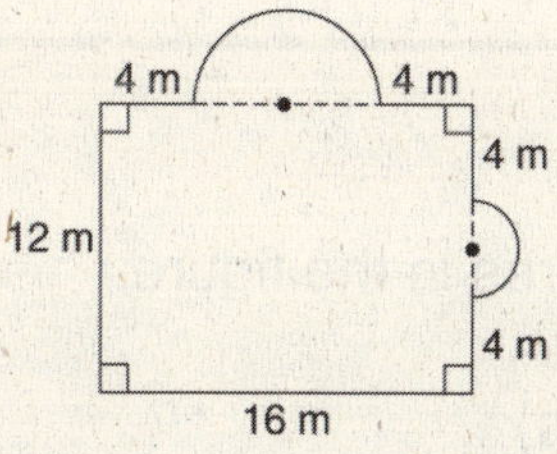

____

5. 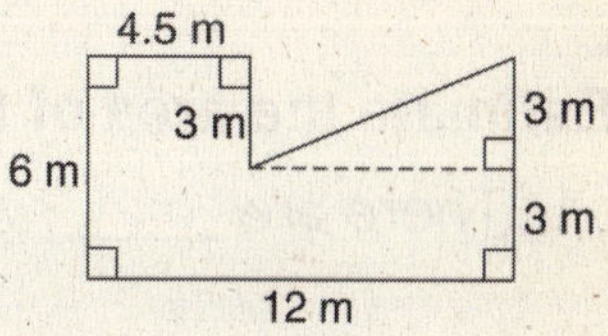

____

6. The figure shows the dimensions of a room in which wedding receptions are held. The room is being carpeted. The three semi-circular parts of the room are congruent. How much carpet is needed?

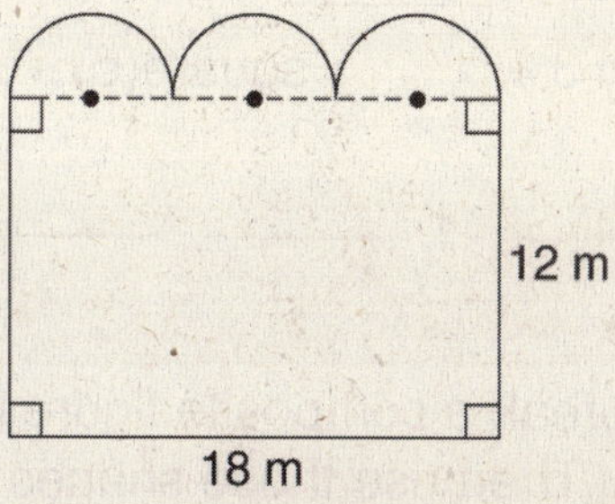

____

7. A polygon has vertices at $F(-5, 2)$, $G(-3, 2)$, $H(-3, 4)$, $J(1, 4)$, $K(1, 1)$, $L(4, 1)$, $M(4, -2)$, $N(6, -2)$, $P(6, -3)$, and $Q(-5, -3)$. Graph the figure on the coordinate plane. Then find the area and the perimeter of the figure.

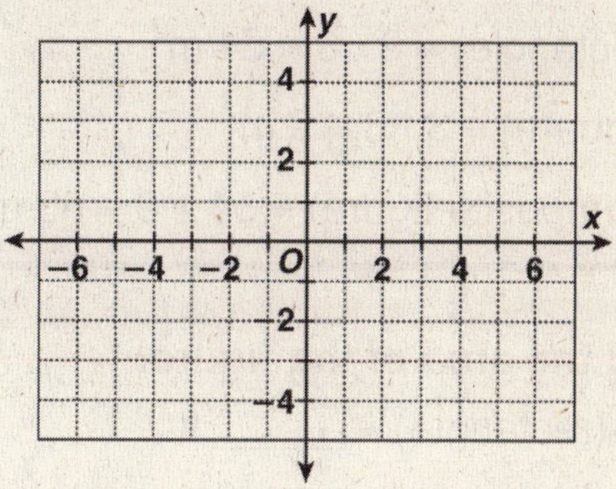

____

Name ______________________ Date ______________ Class ______________

LESSON 9-6

Review for Mastery

Area of Irregular Figures

When an irregular figure is on graph paper, you can estimate its area by counting whole squares and parts of squares. Follow these steps.

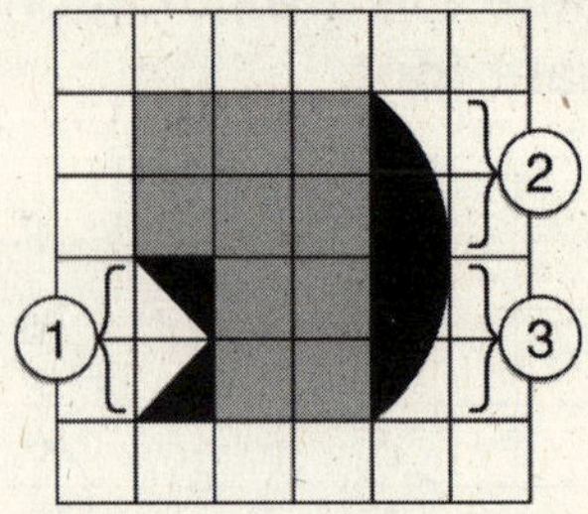

- Count the number of whole squares. There are 10 whole squares.
- Combine parts of squares to make whole squares or $\frac{1}{2}$-squares

Section 1 = 1 square

Section 2 ≈ $1\frac{1}{2}$ squares

Section 3 ≈ $1\frac{1}{2}$ squares

- Add the whole and partial squares.

$10 + 1 + 1\frac{1}{2} + 1\frac{1}{2} = 14$

The area is about 14 square units.

Estimate the area of the figure.

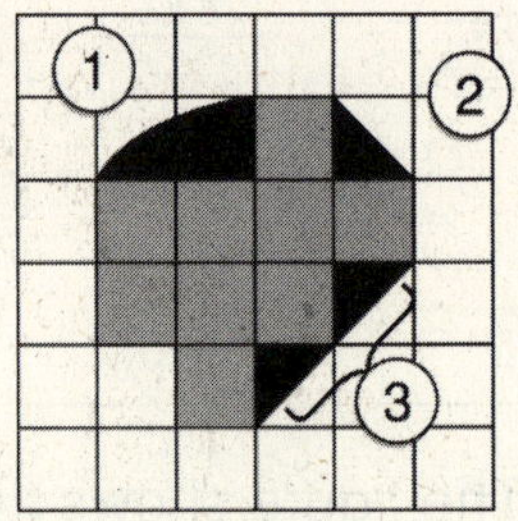

1. There are _____ whole squares in the figure.

Section 1 ≈ _______ square(s)

Section 2 = _______ square(s)

Section 3 = ______ square(s)

A = _______ + _______ + _______ + _______ = _______ square units

You can break a composite figure into shapes that you know. Then use those shapes to find the area.

A (rectangle) = $9 \cdot 6 = 54$ m^2

A (square) = $3 \cdot 3 = 9$ m^2

A (composite figure) = $54 + 9 = 63$ m^2

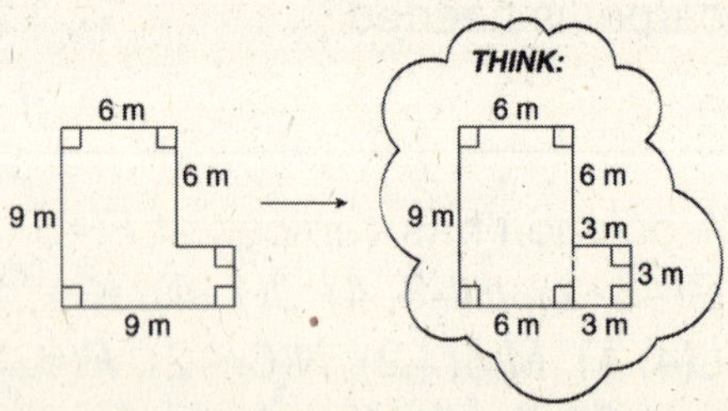

Find the area of the figure.

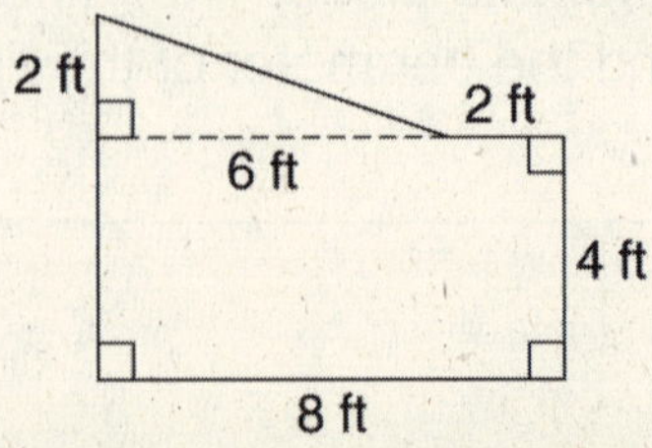

2. *A* (rectangle) = ______ ft^2

A (triangle) = ______ ft^2

A (irregular figure) = ______ + ______

= ______ ft^2

Name ______________________ Date ______________ Class ______________

LESSON 9-6

Challenge

Figure it Out!

Sometimes there is more than one way to find the area of a composite figure.

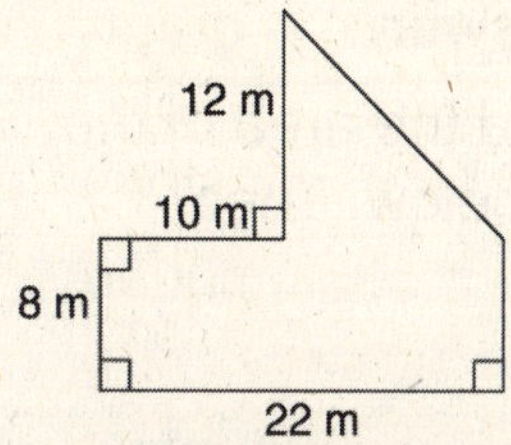

One Way

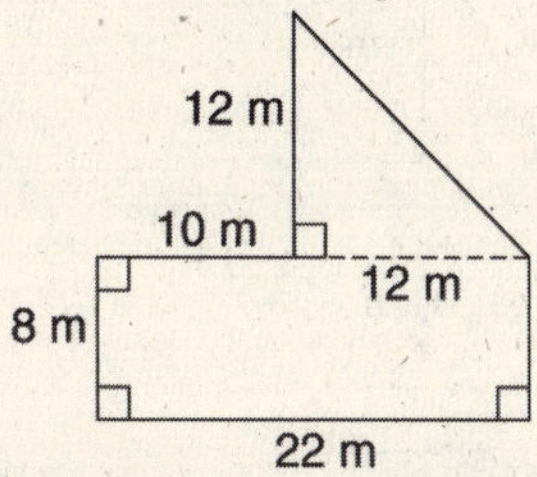

A(rectangle) = $22 \cdot 8 = 176\ \text{ft}^2$

A(triangle) = $\frac{1}{2} \cdot 12 \cdot 12 = 72\ \text{ft}^2$

A(figure) = $176 + 72 = 248\ \text{ft}^2$

Another Way

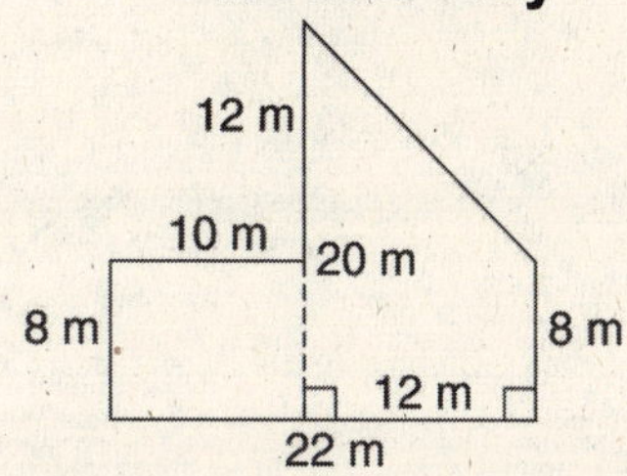

A(rectangle) = $10 \cdot 8 = 80\ \text{ft}^2$

A(trapezoid) = $\frac{1}{2} \cdot 12 \cdot (20 + 8) = 168\ \text{ft}^2$

A(figure) = $80 + 168 = 248\ \text{ft}^2$

Show two different ways to find the area of each figure.

1. **One Way** **Another Way**

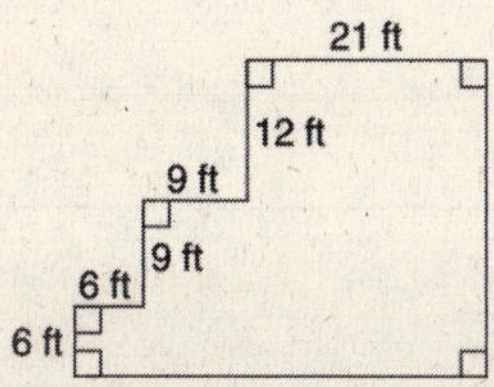

Area = __________ Area = __________

2

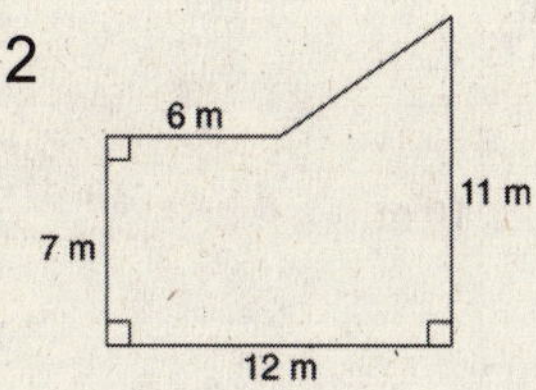

Area = __________ Area = __________

3.

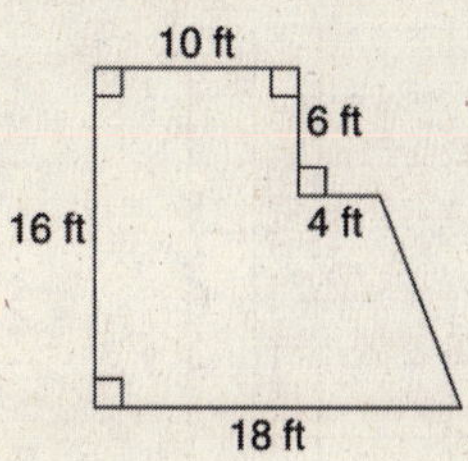

Area = __________ Area = __________

Name ______________________ Date ______________ Class ______________

LESSON 9-6

Problem Solving

Area of Irregular Figures

Write the correct answer.

1. Explain how to find the area of the composite figure below. Then find the area.

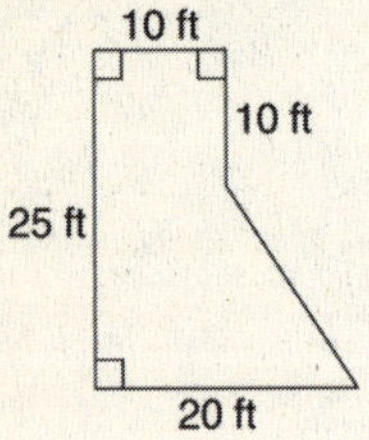

2. Mr. Bemis carpets the living room shown below. If he pays $20 per square meter, what is the total cost of the carpet?

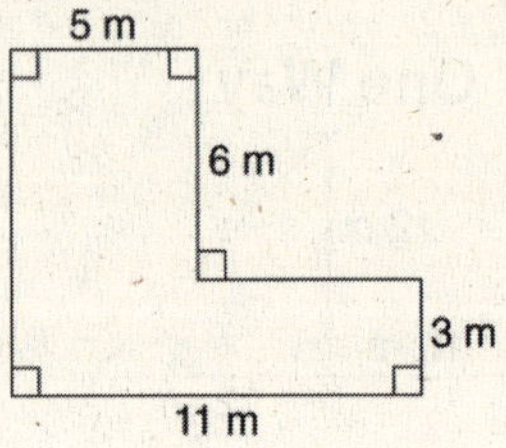

3. A figure is made of a square and a semi-circle. The square has sides of 16 cm each. One side of the square is also the diameter of the semi-circle. What is the total area of the figure? Use 3.14 for π.

4. A figure is made of a rectangle and an isosceles right triangle. The rectangle has sides of 6 in. and 3 in. One of the short sides of the rectangle is also one of the legs of the right triangle. What is the total area of the figure?

Choose the letter of the correct answer.

5. Norene builds the deck at the right. The area of the deck is 10 m^2 greater than was originally planned. What is the area of the deck?

   A 110 m^2 C 66 m^2

   B 76 m^2 D 56 m^2

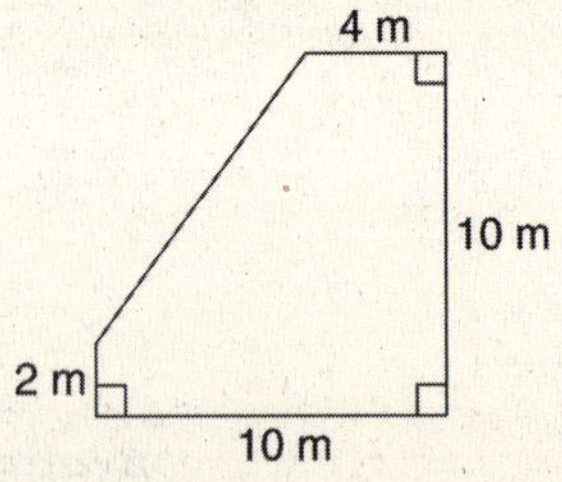

6. The grid to the right shows a swimming pool. Each square represents 1 square meter. What is the best estimate of the area of the swimming pool?

   F 45 m^2 H 37 m^2

   G 41 m^2 J 32 m^2

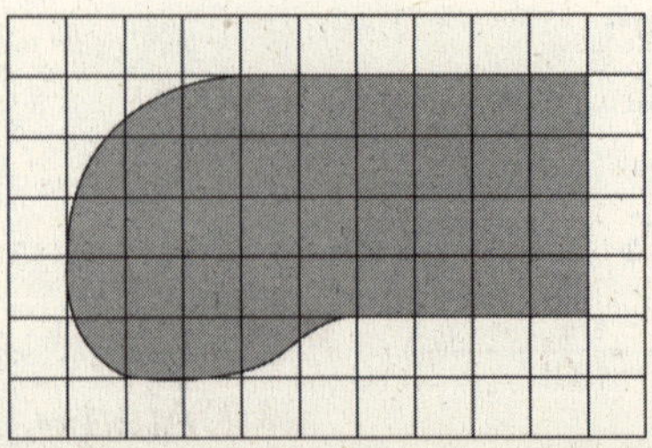

Name ______________________ Date ______________ Class ______________

LESSON 9-6

Reading Strategies

Draw a Picture

A composite figure is made of shapes that you know, such as rectangles, triangles, or semi-circles.

You can draw a line to break a composite figure into other shapes. Then you can use those shapes to help you find the area.

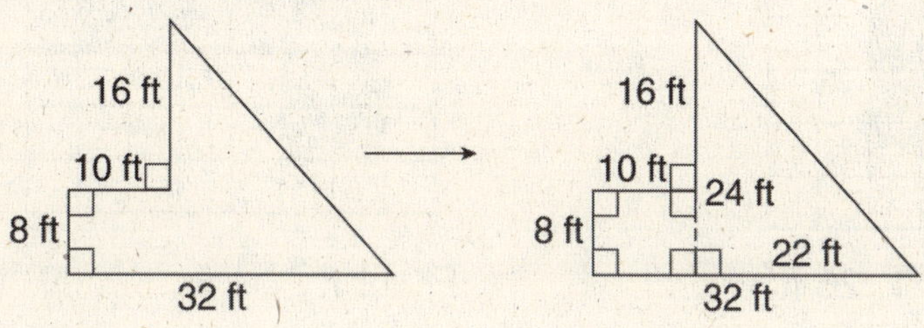

Answer each question.

1. A dashed line has been drawn through the figure. What two shapes does the dashed line create?

____

2. How can you find the area of the rectangle?

____

3. What is the area of the rectangle?

____

4. How can you find the area of the right triangle?

____

5. What is the area of the right triangle?

____

6. How can you use the area of the rectangle and the area of the right triangle to find the total area of the composite figure?

____

7. What is the total area of the composite figure?

____

Name ______________________ Date ______________ Class ______________

LESSON 9-6

Puzzles, Twisters & Teasers

What's Bugging You?

Each square represents 1 square foot. Estimate the area of each irregular figure. Circle the letter next to the better estimate.

1.

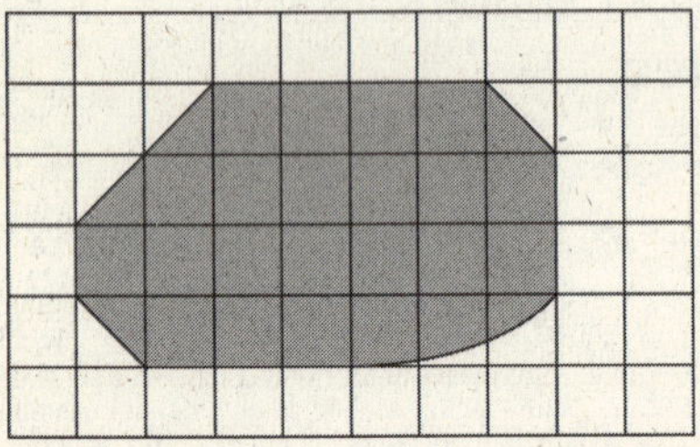

D 24 ft^2 I 26 ft^2

2.

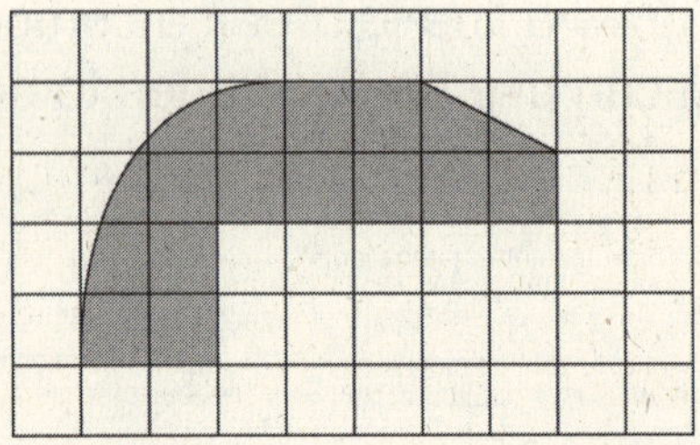

F 11 ft^2 N 15 ft^2

Next, find the area of each composite figure. Use 3.14 for π. Circle the letter next to your answer.

3.

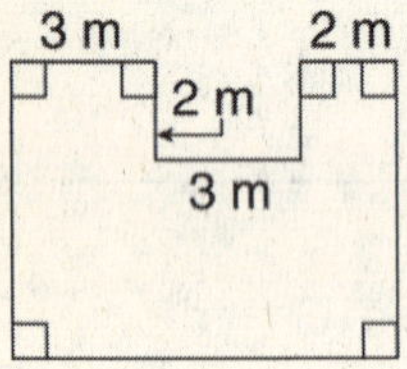

L 38 m^2

R 42 m^2

4.

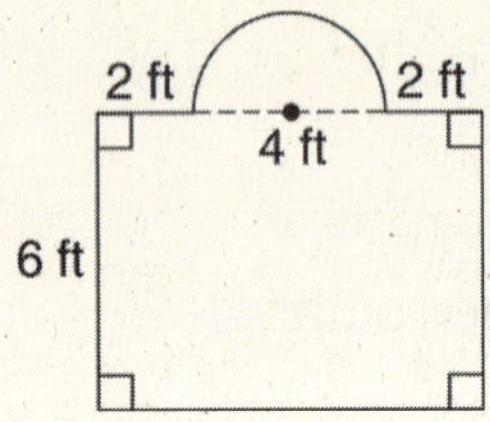

A 54.28 ft^2

E 60.56 ft^2

5.

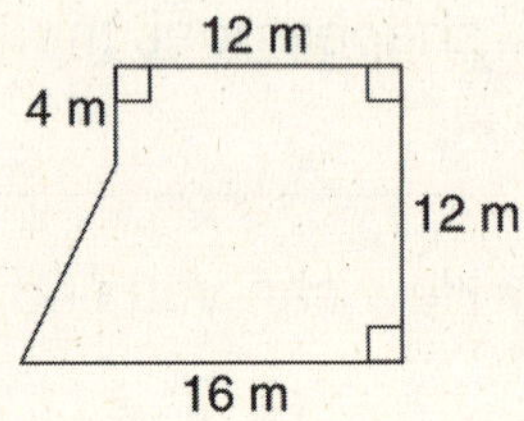

O 160 m^2

E 168 m^2

6.

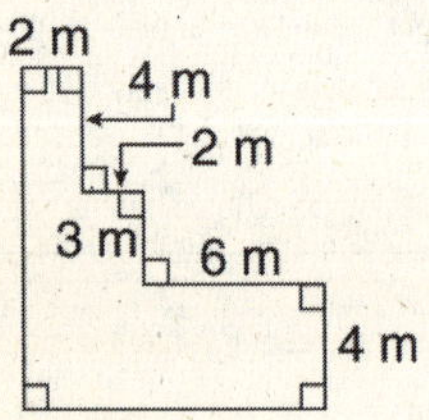

H 52 m^2

L 60 m^2

7.

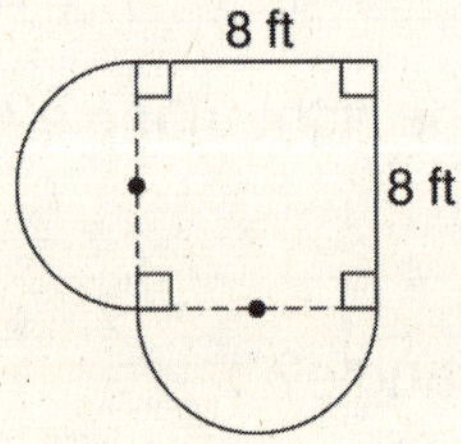

S 89.12 ft^2

Y 114.24 ft^2

8.

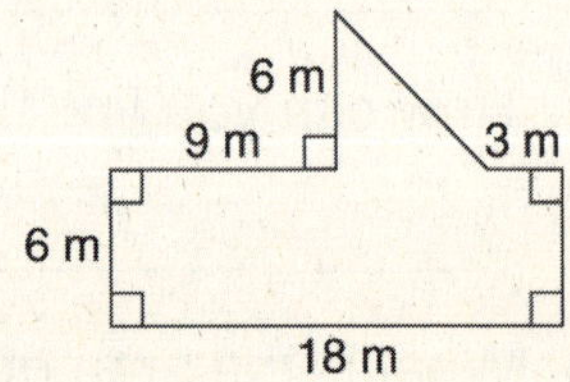

F 126 m^2

B 144 m^2

Write the circled letters above the problem numbers to solve the riddle.

What kind of insect breathes fire?

A ____ ____ ____ G ____ ____ ____ ____ ____

1. 3. 4. 5. 2. 8. 6. 7.

CHAPTER 9

Family Letter

9B Using Squares and Square Roots

Dear Family,

The student is learning to find the square of a number, as well as the square root of a number.

To find the square of a number, the student will multiply the number by itself. A square of a number has a power of 2 and is written in the form x^2, where x is any whole number. To model this concept, the student will use the area formula and a model of a square.

Find each square.

$\mathbf{8^2}$

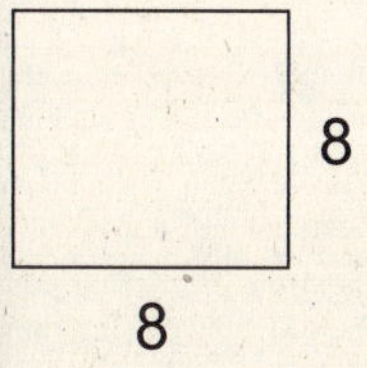

$A = \ell w$

$A = 8 \cdot 8$ Substitute.

$A = 64$ units^2 Multiply.

The square of 8 is 64.

A **square root** is one of the two equal factors of a number.

A number is called a **perfect square** when its square root is a whole number.

The student will estimate the square root of a nonperfect square by determining the two perfect squares the number falls between. He or she can then use a calculator to check the reasonableness of the answer.

Estimate the square root to the nearest whole number. Use a calculator to check your answer.

$\sqrt{40}$

$36 < 40 < 49$ Find the perfect squares closest to 40.

$\sqrt{36} < \sqrt{40} < \sqrt{49}$

$6 < \sqrt{40} < 7$ Find the square roots of 36 and 49.

$\sqrt{40} \approx 6$ 40 is closer to 36 than to 49.

Check $\sqrt{40} = 6.32455532$

The answer is reasonable.

Vocabulary

These are the math words we are learning:

hypotenuse the side opposite the right angle in a right triangle

leg the two sides that form the right angle in a right triangle

perfect square the square of a whole number

Pythagorean Theorem In a right triangle, the sum of the squares of the lengths of the legs is equal to the square of the length of the hypotenuse.

radical sign the symbol for a square root, $\sqrt{\ }$

square root one of the two equal factors of a number

CHAPTER 9

Family Letter

9B Using Squares and Square Roots continued

Calculating squares and square roots is necessary to find the length of a side of a right triangle. A right triangle is a triangle with a 90° angle. The 90° angle in a right triangle is formed by the two **legs** of the triangle. The side opposite the right angle is called the **hypotenuse.**

There is a unique relationship between the length of the legs of a right triangle and the length of the hypotenuse. It is called the **Pythagorean Theorem.** The student will use this theorem in many different problem-solving situations.

Words	Formula	Figure
In a right triangle, the sum of the squares of the lengths of the legs is equal to the square of the length of the hypotenuse.	$a^2 + b^2 = c^2$	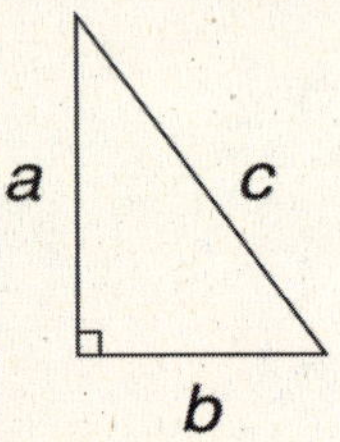

Use the Pythagorean Theorem to find the missing measure.

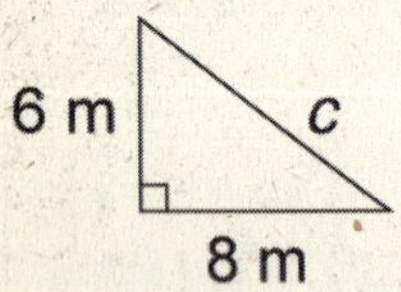

$a^2 + b^2 = c^2$	Use the Pythagorean Theorem.
$6^2 + 8^2 = c^2$	Substitute for *a* and *b*.
$36 + 64 = c^2$	Evaluate the powers.
$100 = c^2$	Add.
$10 = c$	Take the square root of both sides.

The length of the leg is 10 m.

This is just an introduction to the applications of the Pythagorean Theorem. As the student continues the study of mathematics, the concepts learned in this chapter will be the basis for many higher-level math applications.

Sincerely,

Name ______________________ Date ______________ Class ______________

CHAPTER 9

At-Home Practice

9B Using Squares and Square Roots

Find each square.

1. $(8.7)^2$ ______
2. 3^2 ______
3. 16^2 ______

Estimate each square root to the nearest whole number. Use a calculator to check your answers.

4. $\sqrt{27}$ ______
5. $\sqrt{78}$ ______
6. $\sqrt{204}$ ______

7. The area of a square park is 3,000 yd^2. What is the approximate length of each side of the park? Find your answer to the nearest yard.

______

Use the Pythagorean Theorem to find each missing measure.

8. 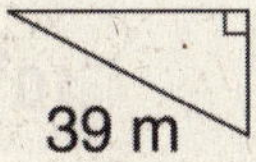

______

9. 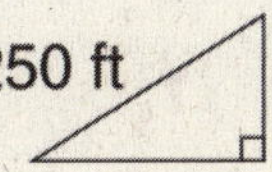

______

10. 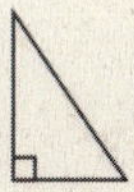

______

Solve.

11. Grace walks 9 blocks south and 7 blocks west to get to her friend's house. How far is Grace from her starting point? Round your answer to the nearest tenth.

______

12. A slide is 20 feet long. If the bottom of the slide is 14 feet from the steps, how tall is the slide? Round your answer to the nearest tenth.

______

Answers: **1.** 8.7 8.7 ; 75.69 **2.** 3 3 ; 9 **3.** 16 16 ; 256 **4.** ≈5; 5.196152423 **5.** ≈9; 8.831760866 **6.** ≈14; 14.28285686 **7.** 55 yd **8.** 36 m **9.** 200 ft **10.** 30 in. **11.** 11.4 blocks **12.** 14.3 ft

Name ______________________ Date ____________ Class ____________

CHAPTER 9

Family Fun

Perfect Root Memory

Directions

- Cut out the cards. Shuffle them and lay them face down in columns and rows.
- Take turns with your opponent. Turn over two cards at a time.
- A match consists of two cards that are equivalent.
- The player with the most matches wins the game.

9	25	49	81	100
3^2	5^2	7^2	9^2	10^2
3	5	7	9	10
$\sqrt{9}$	$\sqrt{25}$	$\sqrt{49}$	$\sqrt{81}$	$\sqrt{100}$
lose a turn	lose a turn	go again	go again	steal a match

Name ______________________ Date ______________ Class ______________

LESSON 9-7

Practice A

Squares and Square Roots

Find each square.

1. 5^2 ______________
2. 10^2 ______________
3. 11^2 ______________
4. 13^2 ______________
5. 7^2 ______________
6. 12^2 ______________

Find each square root.

7. $\sqrt{225}$ ______________
8. $\sqrt{36}$ ______________
9. $\sqrt{81}$ ______________
10. $\sqrt{289}$ ______________
11. $\sqrt{196}$ ______________
12. $\sqrt{484}$ ______________

Match each square root in Column A with the letter of its nearest whole number estimate in Column B. Use a calculator to check your answer.

Column A	Column B	Column A	Column B
13. $\sqrt{38}$ ______	A. 2	19. $\sqrt{84}$ ______	G. 10
14. $\sqrt{19}$ ______	B. 8	20. $\sqrt{46}$ ______	H. 4
15. $\sqrt{66}$ ______	C. 6	21. $\sqrt{95}$ ______	I. 8
16. $\sqrt{54}$ ______	D. 5	22. $\sqrt{10}$ ______	J. 7
17. $\sqrt{29}$ ______	E. 7	23. $\sqrt{15}$ ______	K. 9
18. $\sqrt{5}$ ______	F. 4	24. $\sqrt{71}$ ______	L. 3

25. The area of a checkerboard is 112 in^2. What is the approximate length of each side of the checkerboard? Find your answer to the nearest inch.

____

Name ______________________ Date ______________ Class ______________

LESSON 9-7

Practice B

Squares and Square Roots

Find each square.

1. 6^2 ________ 2. 19^2 ________ 3. 13^2 ________ 4. 4^2 ________

5. 10^2 ________ 6. 14^2 ________ 7. 20^2 ________ 8. 18^2 ________

Find each square root.

9. $\sqrt{289}$ ________ 10. $\sqrt{49}$ ________ 11. $\sqrt{256}$ ________ 12. $\sqrt{81}$ ________

13. $\sqrt{121}$ ________ 14. $\sqrt{625}$ ________ 15. $\sqrt{576}$ ________ 16. $\sqrt{900}$ ________

Estimate each square root to the nearest whole number.
Use a calculator to check your answer.

17. $\sqrt{11}$ ________ 18. $\sqrt{31}$ ________ 19. $\sqrt{98}$ ________ 20. $\sqrt{50}$ ________

21. $\sqrt{152}$ ________ 22. $\sqrt{14}$ ________ 23. $\sqrt{70}$ ________ 24. $\sqrt{28}$ ________

25. $\sqrt{39}$ ________ 26. $\sqrt{193}$ ________ 27. $\sqrt{119}$ ________ 28. $\sqrt{85}$ ________

29. $\sqrt{5}$ ________ 30. $\sqrt{42}$ ________ 31. $\sqrt{75}$ ________ 32. $\sqrt{215}$ ________

33. The area of a square vegetable garden is 75 ft^2. What is the approximate length of each side of the garden? Find your answer to the nearest foot. ________

34. The area of a square computer screen is 138 in^2. What is the approximate length of each side of the screen? Find your answer to the nearest inch. ________

35. Tim broke a square picture window with his baseball. The area of the window is 52 ft^2. What is the approximate width of the window to be replaced? Find your answer to the nearest foot. ________

36. A square tile has an area of 413 cm^2. What is the approximate length of a side of the tile? Find your answer to the nearest centimeter. ________

Name ______________________ Date ______________ Class ______________

LESSON 9-7

Practice C

Squares and Square Roots

Find each square.

1. 18^2 ________
2. $(0.7)^2$ ________
3. 27^2 ________
4. $(3.9)^2$ ________
5. 50^2 ________
6. $(0.31)^2$ ________
7. $\left(\frac{1}{5}\right)^2$ ________
8. $\left(\frac{7}{8}\right)^2$ ________

Find each square root.

9. $\sqrt{784}$ ________
10. $\sqrt{484}$ ________
11. $\sqrt{225}$ ________
12. $\sqrt{961}$ ________

Estimate each square root to the nearest whole number.
Use a calculator to check your answer.

13. $\sqrt{54}$ ________
14. $\sqrt{86}$ ________
15. $\sqrt{12}$ ________
16. $\sqrt{27}$ ________
17. $\sqrt{71}$ ________
18. $\sqrt{15}$ ________
19. $\sqrt{95}$ ________
20. $\sqrt{118}$ ________
21. $\sqrt{149}$ ________
22. $\sqrt{201}$ ________
23. $\sqrt{250}$ ________
24. $\sqrt{175}$ ________
25. $\sqrt{450}$ ________
26. $\sqrt{375}$ ________
27. $\sqrt{600}$ ________
28. $\sqrt{1{,}000}$ ________

Estimate each sum or difference to the nearest whole number.

29. $\sqrt{20} + \sqrt{30}$ ________
30. $\sqrt{50} + \sqrt{10}$ ________
31. $\sqrt{75} - \sqrt{60}$ ________
32. $\sqrt{7} + 7^2$ ________
33. $5^2 - \sqrt{5}$ ________
34. $\sqrt{13} + 13^2$ ________
35. Find the area of a square whose perimeter is 48 in. ________
36. Find the perimeter of a square whose area is 225 yd^2. ________
37. The area of a square rug is 3,000 in^2. What is the approximate length of each side of the rug? Find your answer to the nearest inch. ________

Name ______________________ Date ______________ Class ______________

LESSON 9-7

Review for Mastery

Squares and Square Roots

Recall that the formula for the area of a square is $A = s^2$. A power in which the exponent is 2 is called a *square*.

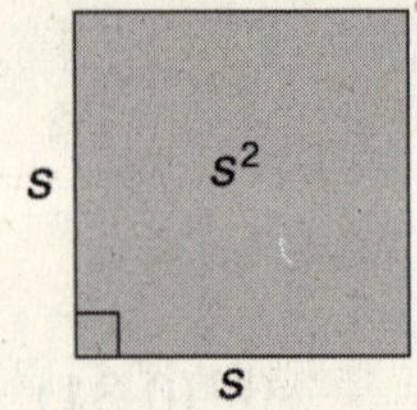

Use the model to help you find the square.

1. 5^2

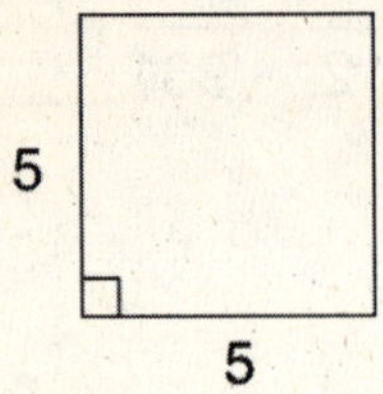

$A = s^2$

$A =$ ______ ^{2}

$A =$ ______ • ______

$A =$ ______

2. 9^2

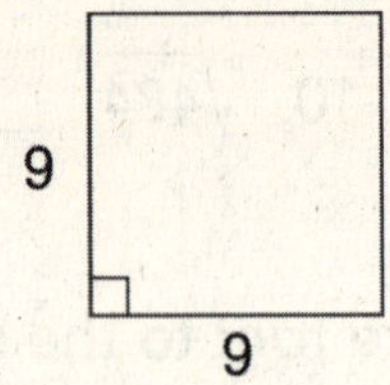

$A = s^2$

$A =$ (______)^2

$A =$ ______ • ______

$A =$ ______

Find the square.

3. 13^2 ______
4. 16^2 ______
5. 11^2 ______
6. 7^2 ______

The numbers 36 and 81 are *perfect squares*. **Perfect squares** are numbers that are the squares of whole numbers.

The square root of 36 is 6. $\sqrt{36} = 6$ because $6 \cdot 6 = 36$.

A table of square roots can help you estimate the square root of a number that is not a perfect square.

Square Root	1	2	3	4	5	6	7	8	9	10
Perfect Squares	1	4	9	16	25	36	49	64	81	100

7. $\sqrt{44}$

Find the perfect square nearest 44.

44 is closest to ______.

Since $\sqrt{49} =$ ______,

$\sqrt{44}$ is closest to ______.

8. $\sqrt{87}$

Find the perfect square nearest 87.

87 is closest to ______.

Since $\sqrt{81} =$ ______,

$\sqrt{87}$ is closest to ______

Name ______________________ Date ______________ Class ______________

LESSON 9-7

Challenge

Power Patterns

1. Complete the tables below.

Powers of 3	
3^1	3
3^2	
3^3	
3^4	
3^5	
3^6	
3^7	
3^8	

Powers of 7	
7^1	7
7^2	
7^3	
7^4	
7^5	
7^6	
7^7	
7^8	

Powers of 9	
9^1	
9^2	
9^3	
9^4	
9^5	
9^6	
9^7	
9^8	

2. Describe the pattern for the ones digits of the powers of 3 values.

____

3. What is the ones digit of 3^9? of 3^{31}? of 3^{98}? ______________

4. Describe the pattern of the ones digits of the powers of 7 values.

____

5. What is the ones digit of 7^{10}? of 7^{24}? of 7^{53}? ______________

6. Do the ones digits for the values of the powers of all whole numbers follow a pattern of *four* repeating digits? Explain.

____

7. What is the ones digit of 11^3? of 11^4? of 11^5? of 11^6? ______________

8. The ones digits of the values of the powers of 11 are the same as the ones digits of the powers of what number? ______________

9. Describe the patterns shown by the powers of 5.

____

10. Describe the pattern of the ones digits for the powers of 6.

____

Name ______________________ Date ______________ Class ______________

LESSON 9-7

Problem Solving

Squares and Square Roots

Write the correct answer. For Problems 1 and 2, use the following formula to find the distance in meters a free-falling object falls from a place of rest: $d = 0.5 \cdot 9.8 \cdot t^2$ (t = time in seconds).

1. As part of a science experiment, Hsing drops a ball from the roof of the school. How far does the ball fall in 2 seconds?

   ____

2. Mel drops a stone from the edge of a cliff overlooking the ocean. How far does the stone fall in 5 seconds?

   ____

3. At the county fair, the apple pies are lined up side-by-side for judging along the length of a 6-foot table. Each pie has an area of 50.24 in^2. How many pies are on the table?

   ____

4. The community swimming pool has an area of 1,024 square feet. The pool is in the shape of a square. What is the perimeter of the pool?

   ____

Choose the letter of the correct answer.

5. The Portuguese national flag is a rectangle. In the center of the flag is a coat of arms and shield on a circle. This circle has a diameter that is half the flag's height. If the flag's circle has an area of 3.14 square feet, what is the height of the flag?

   A 1 ft
   B 3 ft
   C 4 ft
   D 2 ft

6. A square picture has an area of 81 square inches. The perimeter of the frame for the picture is 8 inches longer than the perimeter of the picture itself. What is the length of each side of the square frame for this picture?

   F 8 in.
   G 10 in.
   H 9 in.
   J 11 in.

7. A basketball game starts with a jump ball. This occurs in the center of the basketball court within a circle that has an area of 113.04 square feet. What is the radius of this circle?

   A 3 ft
   B 6 ft
   C 9 ft
   D 36 ft

8. A square fence encloses a vegetable garden with an area of 169 square feet. What is the perimeter of the fence?

   F 52 ft
   G 26 ft
   H 13 ft
   J 56.25 ft

Name ________________________ Date ______________ Class ______________

LESSON 9-7

Reading Strategies

Use A Graphic Organizer

A **square number** is a whole number multiplied by itself.

$3 \times 3 =$ **9** **9** is a square number.

$6 \times 6 =$ **36** **36** is a square number.

A square number has two factors that are the same.

A **square root** is the opposite of a square number.

This chart helps you picture square numbers and square roots.

Perfect Square

4 by 4

$A = \ell\omega = (4 \cdot 4)$

$A = 4^2$

Read: "four squared."

$A = 16$ square units

16 is a square number.

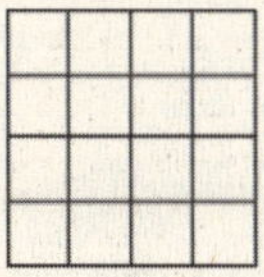

4

4

Not a Perfect Square

4 by 5

$A = \ell\omega = (4 \cdot 5)$

$A = 20$

$A = 20$ square units

20 is not a square number.

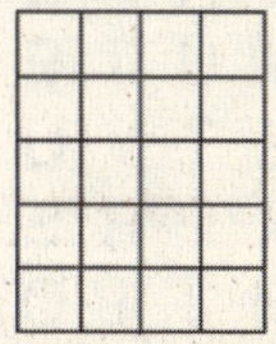

5

4

Squares and Square Roots

Square Root

- The opposite of a square
- $\sqrt{\ }$ is the radical symbol used to write square roots.

$\sqrt{16} = 4$

Read: "The square root of 16 equals 4."

$\sqrt{20} \approx 4.47$

Read: "The square root of 20 equals approximately 4.47."

Use the chart to help you answer each question.

1. How can you tell if a number is a square number?

____

2. Write how to read 5^2?

____

3. Complete: $5^2 =$ ___________________________

4. Write how to read $\sqrt{25}$?

____

Name ______________________ Date ______________ Class ______________

LESSON 9-7

Puzzles, Twisters & Teasers

The Root of the Matter!

Find and circle the words in the box below in the word search. The words may be horizontal or vertical. Find a word in the word search that solves the riddle. Circle it and write it on the line.

power	root	perfect	square	radical
sign	express	value	symbol	model

```
P E R F E C T R C V B
O V U J E R V A L U E
W G T Y U I L D S Q W
E X P R E S S I Y E R
R M J I O Q I C M Z X
O V B N V U G A B O I
O I O P L A N L O P U
T B H V I R U S L L I
S A M O D E L A S E C
```

Why did the computer sneeze?

It had a ______________________.

Name ______________________ Date ______________ Class ______________

LESSON 9-8

Practice A

The Pythagorean Theorem

Use the Pythagorean Theorem to find each missing measure. Choose the letter for the best answer.

1.

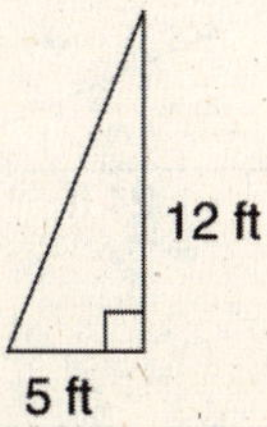

A 7 ft　　C 14 ft
B 13 ft　　D 17 ft

2.

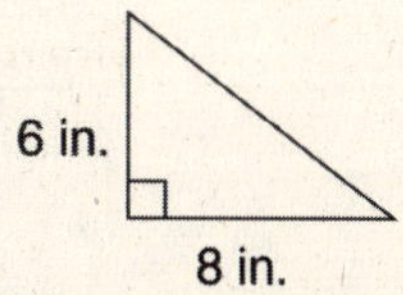

F 14 in.　　H 5 in.
G 10 in.　　J 2 in.

3.

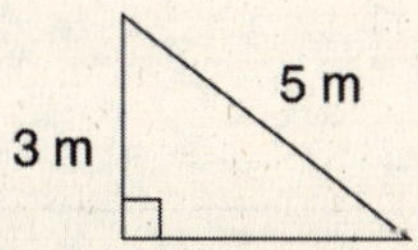

A 2 m　　C 6 m
B 4 m　　D 8 m

4.

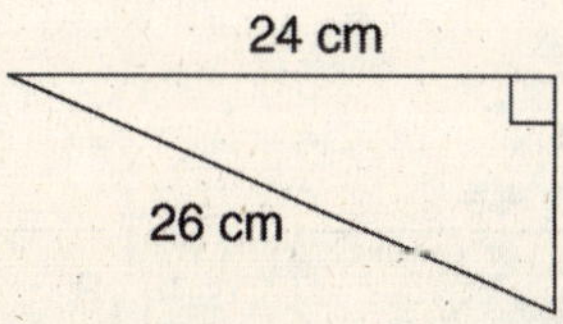

F 2 cm　　H 10 cm
G 4 cm　　J 40 cm

5.

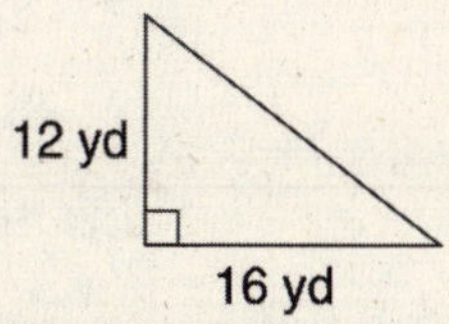

A 11 yd　　C 20 yd
B 16 yd　　D 28 yd

6.

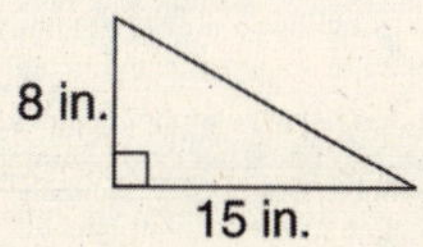

F 23 in.　　H 13 in.
G 17 in.　　J 7 in.

7. A 15-foot ladder is leaning against a wall. The ladder is 6 ft from the base of the wall. About how far above the ground does the ladder touch the wall? Round your answer to the nearest tenth.

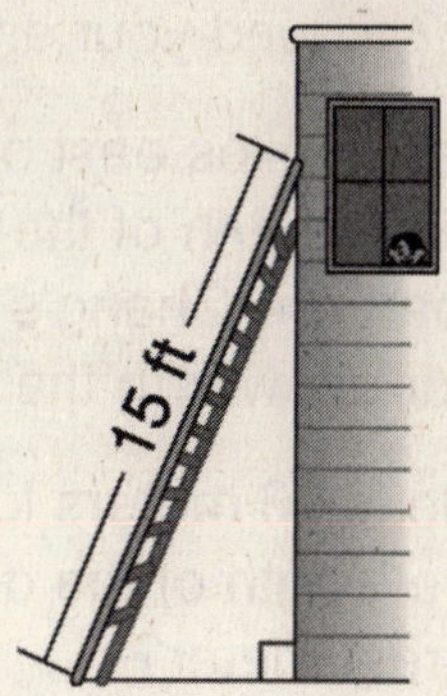

____

Name ______________________ Date ____________ Class ____________

LESSON 9-8

Practice B

The Pythagorean Theorem

Use the Pythagorean Theorem to find each missing measure.

1.

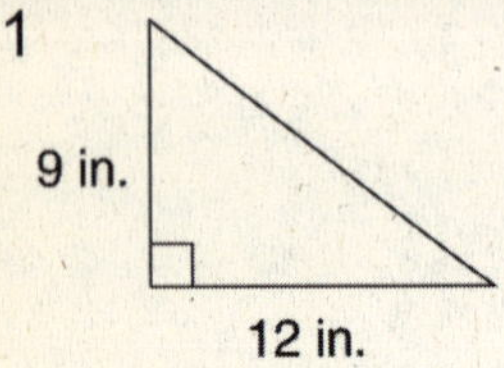

2.

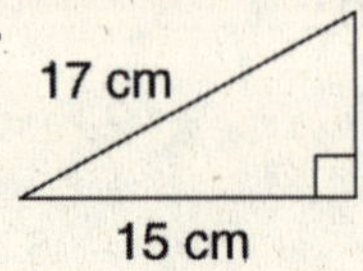

3.

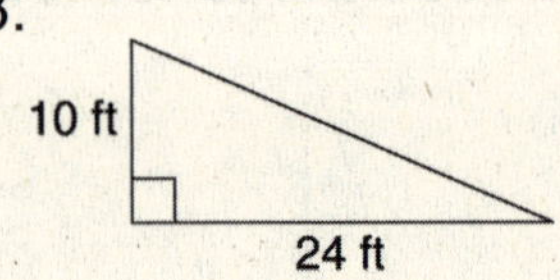

4.

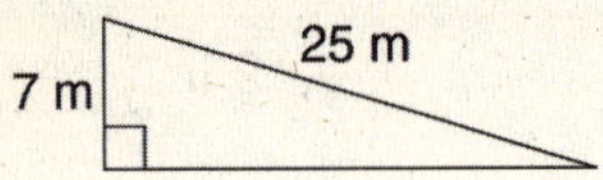

5.

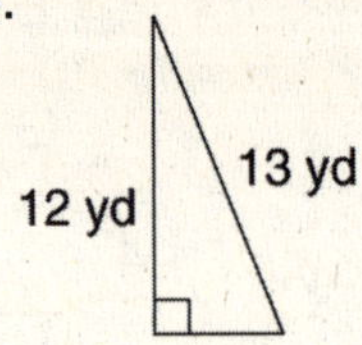

6.

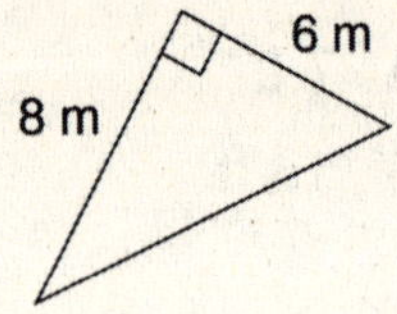

7.

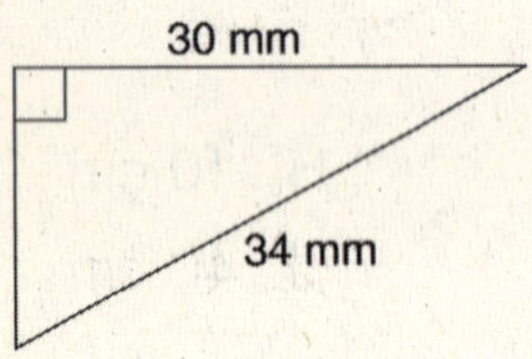

8.

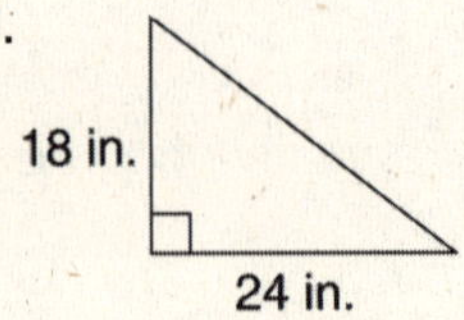

9.

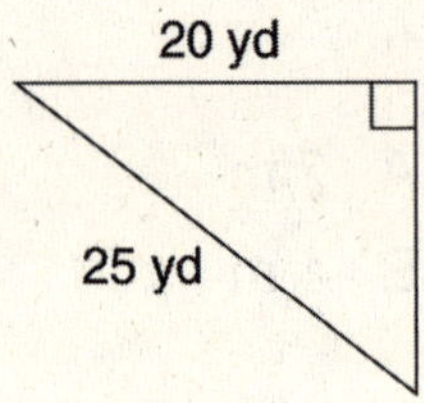

10. A 20-ft ladder is leaning against a wall. If the ladder is 12 ft from the base of the wall, how high above the ground does the ladder touch the wall? ____________

11. A checkerboard is 10 inches long on each side. What is the length of the diagonal from one corner to another? Round your answer to the nearest tenth. ____________

12. Chang lives 8 miles east of the school. Deborah lives 12 miles south of the school. Approximately how far apart are Chang's and Deborah's homes? Round your answer to the nearest tenth. ____________

13. A rectangle is 26 meters long and 18 meters wide. What is the length of the diagonal of the rectangle to the nearest meter? ____________

Name ______________________ Date ______________ Class ______________

LESSON 9-8

Practice C

The Pythagorean Theorem

Use the Pythagorean Theorem to find each missing measure.

1. 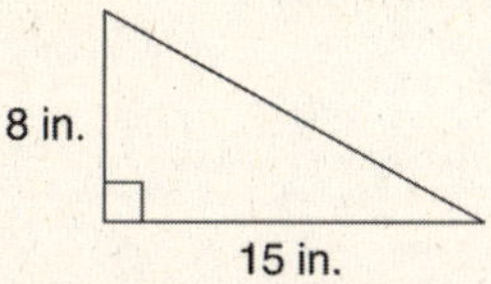

____

2. 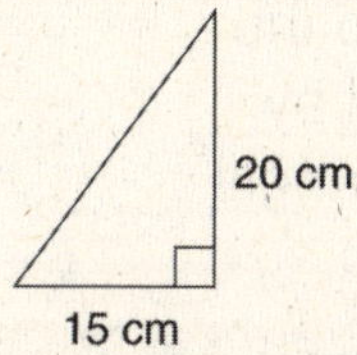

____

3. 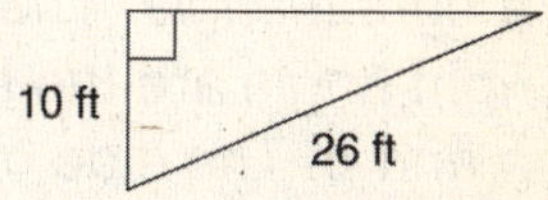

____

4. 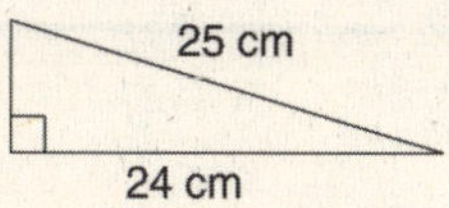

____

5. 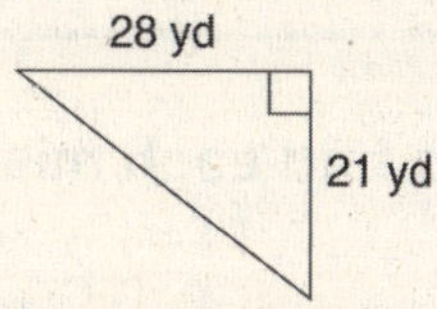

____

6. 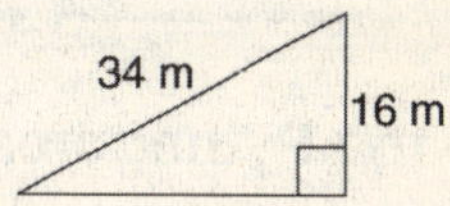

____

The lengths of two sides of a right triangle are given. Find the length of the third side to the nearest tenth.

7. legs: 4 ft and 9 ft ______________

8. leg: 8 m; hypotenuse: 12 m ______________

9. leg: 15 ft; hypotenuse: 20 ft ______________

10. legs: 10 m and 7 m ______________

Determine whether each set of lengths forms a right triangle.

11. $a = 10$, $b = 24$, $c = 25$ ______

12. $a = 12$, $b = 16$, $c = 20$ ______

13. $a = 4$, $b = 6$, $c = 8$ ______

14. $a = 5$, $b = 12$, $c = 13$ ______

Solve.

15. The foot of a ladder is 8 ft from the base of a wall. If the ladder is 20 ft long, how high up on the wall does the ladder reach? Round your answer to the nearest tenth. ______________

16. The hypotenuse of an isosceles right triangle is 16 centimeters. Find the length of the two other sides of the triangle to the nearest tenth. ______________

Name ______________________ Date ______________ Class ______________

LESSON **9-8**

Review for Mastery

The Pythagorean Theorem

A triangle containing a right angle is called a *right triangle.* The sides adjacent to the right angle are the **legs,** represented by *a* and *b*. The side opposite the right angle is the **hypotenuse,** represented by *c*.

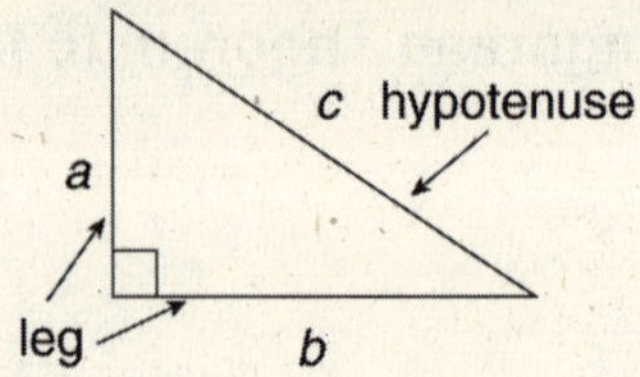

The **Pythagorean Theorem** states:

If *a* and *b* are the lengths of the legs of a right triangle and *c* is the length of the hypotenuse, then $a^2 + b^2 = c^2$.

Use the Pythagorean Theorem to find each missing length.

1.

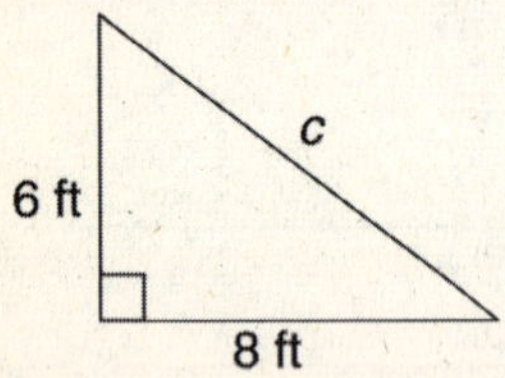

$a^2 + b^2 = c^2$

_____^{2} + _____^{2} = c^2

_____ + _____ = c^2

_______ = c^2

c = _______

2.

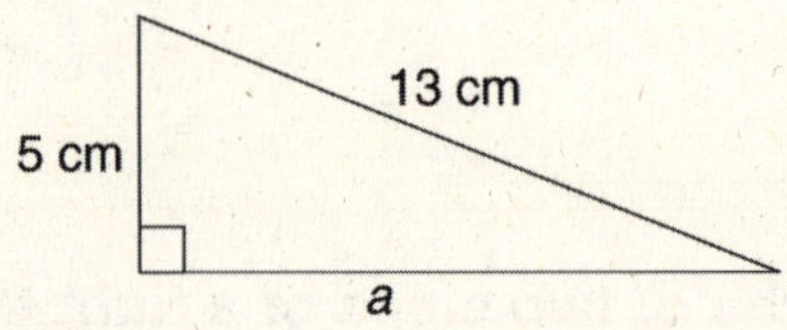

$a^2 + b^2 = c^2$

a^2 + _____^{2} + _____^{2}

a^2 + _____ + _____

a^2 = _____

a = _______

3.

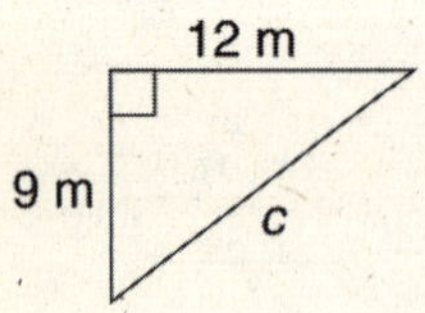

____

4.

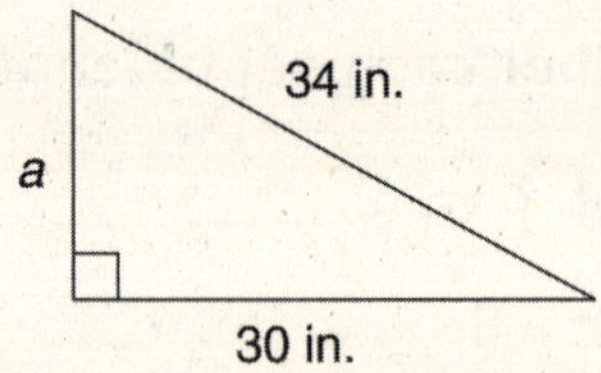

____

5.

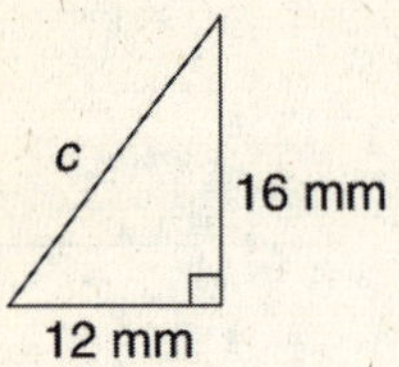

____

6.

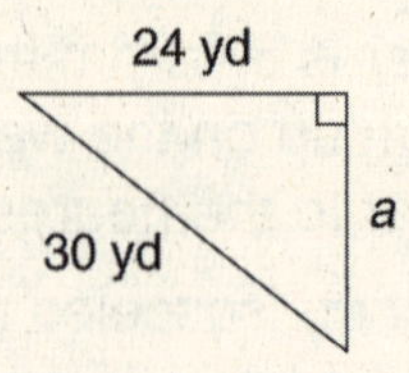

____

Name ______________________ Date ______________ Class ______________

LESSON 9-8

Challenge

Triangle Families

Any three numbers that satisfy the formula $a^2 + b^2 = c^2$ form a *Pythagorean triple.*

$3^2 + 4^2 = 5^2$

$9 + 16 = 25$

(3, 4, 5) is a Pythagorean triple.

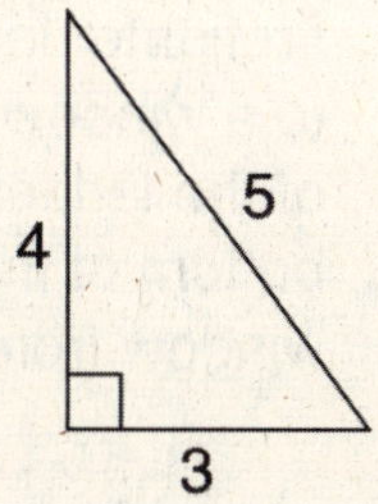

When you multiply or divide a Pythagorean triple by the same number, you get another Pythagorean triple.

So, (3 • 2, 4 • 2, 5 • 2),

or (6, 8, 10),

is a Pythagorean triple.

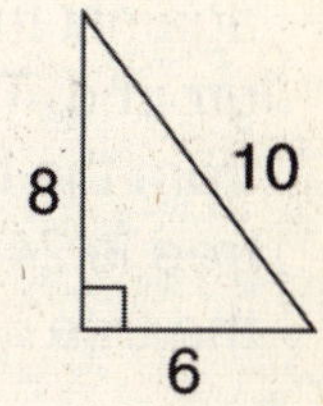

Check it out: $6^2 + 8^2 = 10^2$

$36 + 64 = 100$

Complete each table to find new Pythagorean triples. Then check some of your results.

1. **The (3, 4, 5) Family**

Multiply by 3.	(9, 12, 15)
Multiply by 5.	
Multiply by 10.	
Divide by 5.	

2. **The (5, 12, 13) Family**

Multiply by 2.	
Multiply by 3.	
Multiply by 8.	
Divide by 10.	

3. **The (7, 24, 25) Family**

Divide by 4.	
Multiply by 2.	
Multiply by 4.	
Multiply by 12.	

4. **The (8, 15, 17) Family**

Divide by 2.	
Multiply by 3.	
Multiply by 4.	
Multiply by 20.	

5. Use the formula to check one triple of the (3, 4, 5) family.

____

6. Use the formula to check one triple of the (5, 12, 13) family.

____

7. Use the formula to check one triple of the (7, 24, 25) family.

____

Name ______________________ Date ______________ Class ______________

LESSON 9-8

Problem Solving

The Pythagorean Theorem

Write the correct answer.

1. During a storm, a tree falls toward a house. The top of the tree leans against the house 45 feet above the ground. The distance on the ground from the house to the base of the tree is 24 feet. What is the height of the tree?

____

2. During a training exercise, a firefighter leans a 40-foot ladder up to a window in a house. The bottom of the ladder is 24 feet from the bottom of the house. How high is the window from the ground?

____

3. A triangle has a hypotenuse of 25 centimeters and a base of 20 centimeters. What is the area of this right triangle?

____

4. The football field at the University of Texas at Arlington is 60 yards by 100 yards. Is the length of the diagonal across this field more or less than 200 yards? Explain.

____

____

Choose the letter of the correct answer.

5. The minimum size of a soccer field for players under 8 years of age is 20 yards by 30 yards. About how far is the diagonal distance on a field with these dimensions?

A about 12 yd
B about 25 yd
C about 36 yd
D about 45 yd

6. The minimum size of a soccer field for international matches is 70 yards by 110 yards. If a player runs diagonally across this field, about how much farther does she run than if the field were 50 yards by 100 yards?

F about 242 yd
G about 130 yd
H about 112 yd
J about 19 yd

7. In the state of Virginia, Winchester is 21 miles north of Front Royal. Arlington is 58 miles east of Front Royal. What is the distance to the nearest mile from Winchester to Arlington?

A 37 mi
B 62 mi
C 89 mi
D 441 mi

8. On a child's slide, the distance from the bottom rung to the top of the ladder is 6 feet. The straight distance from the bottom rung of the ladder to the bottom of the slide is 36 inches. About how long is the slide?

F about 6.7 ft
G about 8.4 ft
H about 9.0 ft
J about 36.5 ft

Name ______________________ Date ______________ Class ______________

LESSON 9-8

Reading Strategies

Understanding Vocabulary

A **right triangle** has special features.

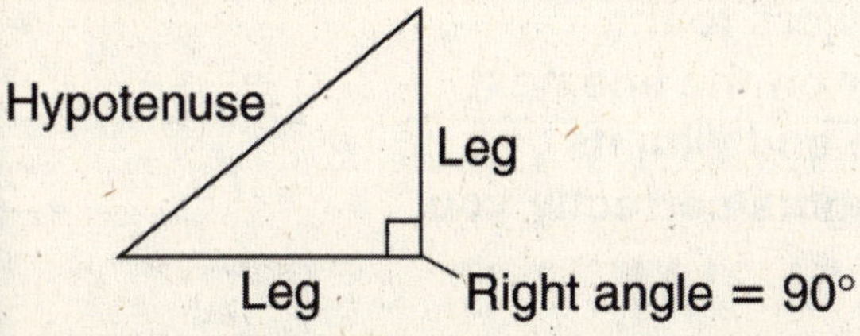

Use the triangle to complete each sentence.

1. A right triangle has one right angle that measures ______________ degrees.
2. The two shorter sides of a right angle are called ______________.
3. The longest side of a right triangle is called the ______________.

The **Pythagorean Theorem** describes a special relationship among the sides of a right triangle. The theorem states that the sum of the legs squared is equal to the hypotenuse squared.

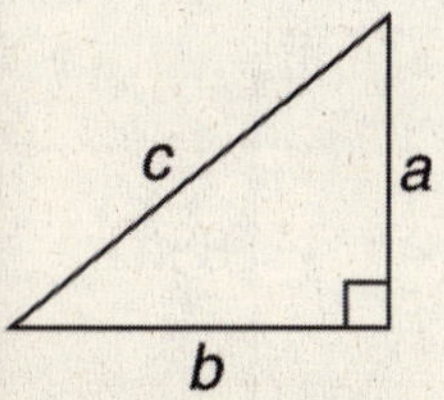

$a^2 + b^2 = c^2$

Look at the figure of the triangle to answer each question.

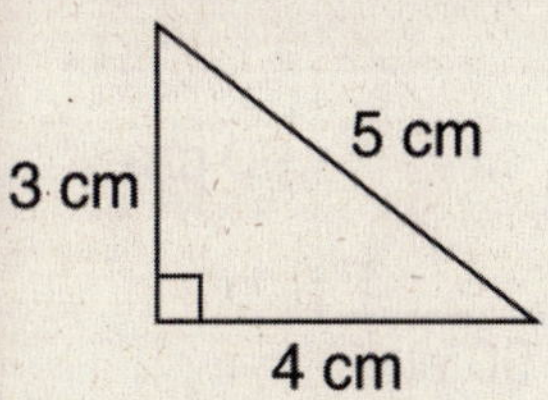

4. What is 3 cm squared? ______________
5. What is 4 cm squared? ______________
6. What is the total of the squares of the legs? ______________
7. What is 5 cm squared? ______________
8. What do you notice about the sum of the squares of the legs and the square of the hypotenuse? ______________

Name ______________________ Date ______________ Class ______________

LESSON 9-8

Puzzles, Twisters & Teasers

Fair and Square!

Use the Pythagorean Theorem to find the length of the hypotenuse of each triangle. Round your answers to the nearest tenth, if necessary. Write your answer on the line next to the letter. Replace the answers listed at the end with its corresponding letter. If you calculated the lengths correctly you will solve the riddle!

1. ______________ = **K**

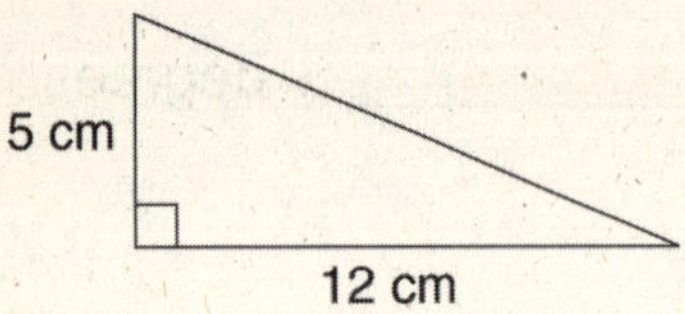

2. ______________ = **E**

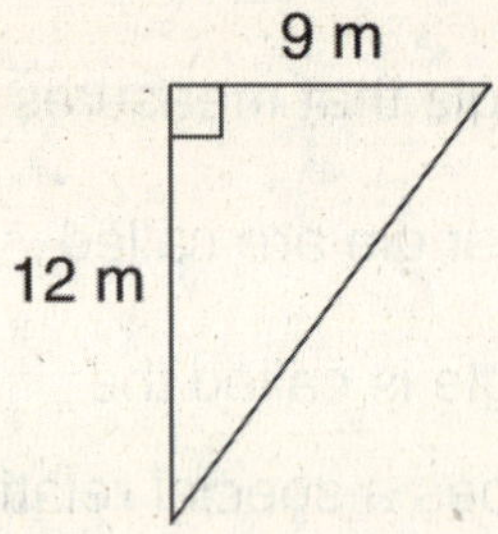

3. ______________ = **N**

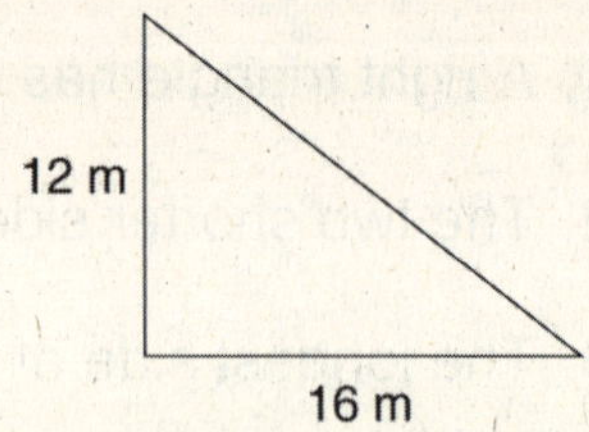

4. ______________ = **B**

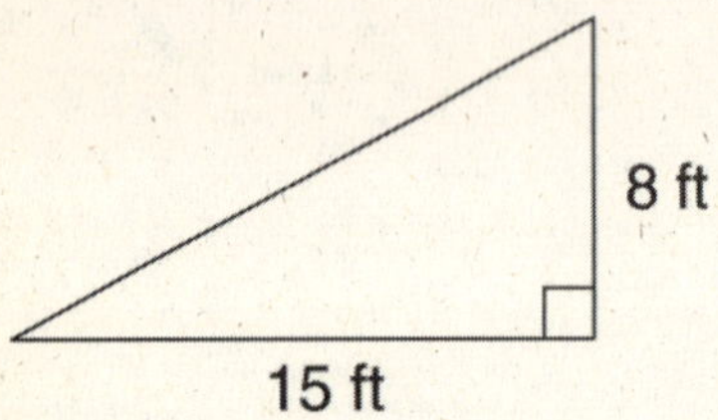

5. ______________ = **U**

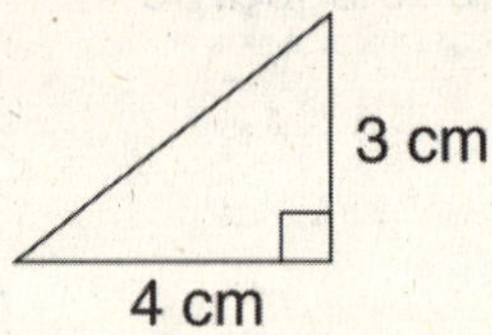

6. ______________ = **I**

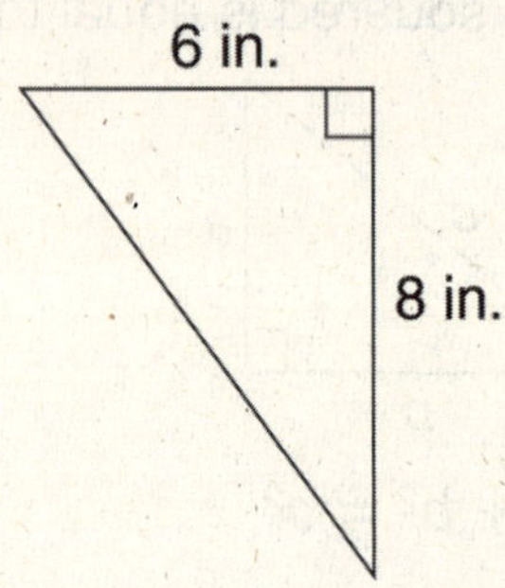

7. ______________ = **R**

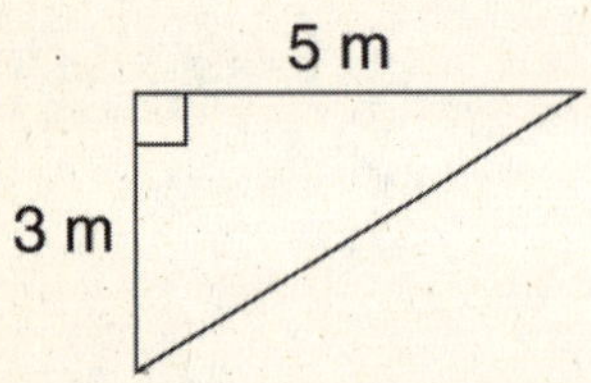

8. ______________ = **G**

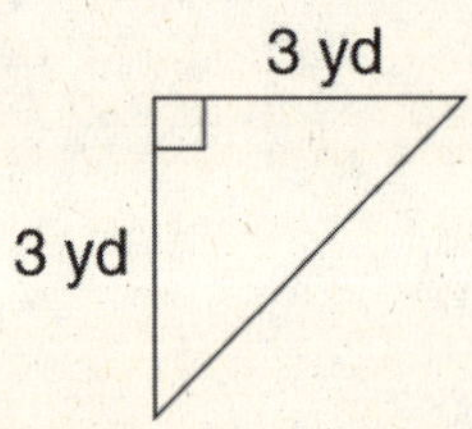

9. ______________ = **T**

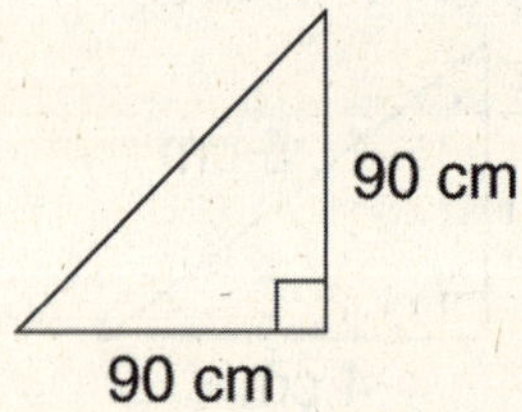

Why did the woman want an elephant instead of a car?

It had a ___ ___ ___ ___ ___ ___ ___ ___ ___ ___ ___.

17 10 4.2 4.2 15 5.8 127.3 5.8 5 20 13

Answers

LESSON 9-1

Practice A

1. 1,200 yd
2. 84.2 oz
3. 200 km
4. 12 fl oz
5. $11\frac{1}{2}$ pt
6. 25 ft
7. 2 in.; 2 in.; $2\frac{1}{8}$ in.; All measurements are equally accurate.
8. 3 in.; $3\frac{1}{4}$ in.; $3\frac{1}{8}$ in.; $3\frac{1}{8}$ in.
9. about 33 in.
10. about $12\frac{1}{2}$ lb
11. 15.9 inches, because an inch is a smaller unit than a foot
12. 0.11 inches, because it is carried out to the hundredths place

Practice B

1. 1,230 m
2. 91.1 kg
3. 14 mm
4. 5,281 in.
5. 1,000.3 mi
6. 550 g
7. $2\frac{1}{2}$ in.; $2\frac{1}{2}$ in.; $2\frac{5}{8}$ in.; All measurements are equally accurate.
8. 2 in.; 2 in.; $2\frac{1}{8}$ in.; All measurements are equally accurate.
9. about 3 kg
10. about 165° F
11. 4.13 inches, because an inch is a smaller unit than a foot
12. 16.67 inches, because it is carried out to the hundredths place

Practice C

1. 1,500 m
2. 90.4 °F
3. 18.75 mi
4. 17 oz
5. 1.11 qt
6. 3,451 mm
7. $2\frac{1}{2}$ in.; $2\frac{1}{2}$ in.; $2\frac{3}{8}$ in.; $2\frac{3}{8}$ in.
8. 2 in.; $2\frac{1}{4}$ in.; $2\frac{1}{8}$ in.; $2\frac{1}{8}$ in.
9. about 6 ft 4 in.
10. about 35°
11. 201.4 inches, because one tenth of an inch is smaller than one tenth of a foot
12. Neither, both are carried out to the tenths

Review for Mastery

1. 4 in.
2. 2 mm
3. 2.1 mm
4. $2\frac{1}{2}$ in, $2\frac{3}{4}$ in; $2\frac{5}{8}$ in; $2\frac{5}{8}$ in.

Challenge

1. –8.8%
2. –0.7125%
3. –0.265%
4. $-0.708\overline{3}\%$
5. 0.74%

Problem Solving

1. centimeters, because a centimeter is smaller than an inch.
2. 7.77 inches, because an inch is smaller than a foot
3. 6,529 feet, because a foot is smaller than a meter
4. 13.95 inches, because an inch is smaller than a foot
5. B
6. A
7. B

Reading Strategies

1. measurement in inches
2. measurement in yards
3. $2\frac{5}{8}$ in.

Puzzles, Twisters & Teasers

1. 43 mi, 17.1 mi, 1.36 mi, 7 ft, 2.3 ft, 1.11 ft, 1.2 in., 1.09 in.; 1.09
2. 4 km, 4.9 km, 99 m, 28 cm, 1 mm, 1.4 mm; 1.4
3. 3 gal, 44 qt, 9 pt, 53.4 qt, 1 fl oz, 8.5 fl oz, 5 tbsp, 0.6 tsp; 0.6
4. 100 kg, 501.2 kg, 6,000 g, 6,000.3 g, 111 mg, 0.2 mg, 0.05 mg; 0.05

   The total is 3.14 (the approximate value of pi).

LESSON 9-2

Practice A

1. 24 ft
2. 28 in.
3. 36 cm
4. 40 yd
5. 18 m
6. 32 in.
7. C
8. J
9. B
10. F
11. 22 in.

Practice B

1. 43 in.
2. 45 cm
3. 30 ft
4. 42 mm
5. 27 m
6. 20 yd
7. 18.8 in.
8. 56.5 cm
9. 4.7 ft
10. 66 ft
11. 132 mm
12. 88 cm

Practice C

1. 14.2 cm
2. 25 in.
3. 68.2 ft
4. 130 yd
5. 52.2 m
6. $24\frac{1}{2}$ in.
7. 29 in.; 91.1 in.
8. 5.7 yd; 11.3 yd
9. 3.8 cm; 23.9 cm
10. 1.8 m; 3.6 m
11. 9.55 in.

Review for Mastery

1. 11; 13; 15; 39
2. 21; 23; 24; 30; 98
3. 16; 8: 48
4. 9; 28.26; 28.3
5. 13; 26; 81.64; 81.6
6. 40.8 cm
7. 31.4 ft
8. 9.4 in.

Challenge

1. $P = 26m$
2. $P = 2s + 6b$
3. $P = 4\ell + 4\omega$
4. $P = 3\left(\frac{1}{2}\pi d + d\right)$
5. 104 in.
6. 40.2 m
7. 8 in.
8. 77.1 cm

Problem Solving

1. 552 ft
2. 1,080 ft
3. 56.5 in.
4. 15.7 in.
5. 31 in.
6. 44 in.
7. A
8. G
9. C
10. H

Reading Strategies

1. 4, 13, and 10
2. Add the lengths of the three sides.
3. $C = 2\pi r$
4. $2\ell + 2\omega$
5. $C = \pi d$

Puzzles, Twisters & Teasers

1. 44 m
2. 19.8 in.
3. 10 ft
4. 25.1 cm
5. 19.5 cm
6. 18 m
7. 24 in.
8. 30 ft

9. 9.4 ft

A VICIOUS CIRCLE

LESSON 9-3

Practice A

1. C
2. G
3. 62.5 cm^2
4. 36 cm^2
5. 48 cm^2
6. 72 cm^2
7. 36 cm^2
8. $37\frac{1}{2}$ cm^2
9. 12 ft
10. 4,340 in^2

Practice B

1. 104 ft^2
2. $138\frac{3}{4}$ m^2
3. 100.01 in^2
4. 63 yd^2
5. 132 cm^2
6. 41.76 ft^2
7. 15 m^2
8. 35 in^2
9. 15.68 m^2
10. 100.75 cm^2
11. 17.5 ft

Practice C

1. 124.8 yd^2
2. 18 in^2
3. 46 ft^2
4. 135 cm^2
5. 5.04 m^2
6. 21 yd^2
7. 435 in^2
8. 28.8 m^2
9. 40 $units^2$
10. 18 $units^2$
11. 30 $units^2$
12. 35 $units^2$
13. 8 cm
14. $456\frac{3}{4}$ in^2

Review for Mastery

1. 9; 7; 63; 63
2. 12; 5; 60; 60
3. 8; 4; 32; 32
4. 6; 9; 54; 54

Challenge

1. 12,600 ft^2; 540 ft
2. 21,600 ft^2; 690 ft
3. 13,950 ft^2; 540 ft
4. 23,400 ft^2; 690 ft
5. 71,550 ft^2
6. 2,460 ft

Problem Solving

1. 6.1 in.
2. 7.6 ft^2
3. 26 bags of seed
4. 8.5 ft high
5. D
6. F
7. B
8. H

Reading Strategies

1. the width
2. the length
3. Possible answer: You multiply either the length or base times the height or width.
4. A rectangle has length and width; a parallelogram has base and height.

Puzzles, Twisters & Teasers

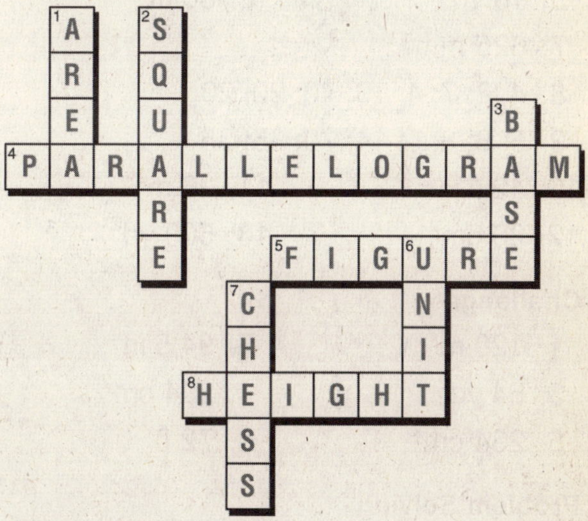

LESSON 9-4

Practice A

1. 21 square units
2. 18 square units
3. 32.5 square units
4. B
5. H
6. A
7. G
8. 69,600 mi^2

Practice B

1. 22 square units
2. 20 square units
3. 31.5 square units
4. 6 square units

5. 49.5 square units
6. 30 square units
7. 21 in^2
8. 103.5 cm^2
9. 63 yd^2
10. 125 in^2
11. 217.5 m^2
12. 65 ft^2
13. 146,450 mi^2

Practice C

1. 66.5 square units
2. 176 square units
3. 77 square units
4. 23.4 ft^2
5. 57.5 cm^2
6. 26.52 in^2
7. 8 yd
8. 2 in^2
9. 46 cm
10. 24 $units^2$
11. 12.5 $units^2$
12. 10,400 mi^2

Review for Mastery

1. 5; 2; 10; 5; 5
2. 9; 8; 72; 36; 36
3. 21 in^2
4. 10.5 cm^2
5. 18 yd^2
6. 80 m^2
7. 625 mm^2
8. 4; 3; 7; 4; 10; 40; 20; 20
9. 5; 6; 8; 5; 14; 70; 35; 35
10. 42 ft^2
11. 36 m^2
12. 20 cm^2
13. 500 yd^2

Challenge

1. 120 m^2
2. 94.5 in^2
3. 64 yd^2
4. 224 cm^2
5. 234 cm^2
6. 72 ft^2

Problem Solving

1. about 6 ft^2
2. You would need 4 times as much, or about 24 ft^2.
3. 3 packets
4. 49.5 ft^2 larger
5. B
6. H

Reading Strategies

1. 9
2. 3
3. 3
4. $4\frac{1}{2}$
5. 8
6. 4
7. The area of a triangle is half the area of a parallelogram or square.

Puzzles, Twisters and Teasers

1. 28
2. 6
3. 15
4. 32
5. 187
6. 25
7. 64
8. 20
9. 54

M I C R O W A V E S

LESSON 9-5

Practice A

1. 50.2 in^2
2. 153.9 m^2
3. 254.3 yd^2
4. 3.1 cm^2
5. 171.9 cm^2
6. 78.5 in^2
7. 379.9 mm^2
8. 19.6 ft^2
9. 28.3 m^2
10. 616 in^2
11. 56.5 in^2

Practice B

1. 113.0 m^2
2. 50.2 ft^2
3. 153.9 yd^2
4. 19.6 cm^2
5. 379.9 in^2
6. 28.3 mm^2
7. 78.5 in^2
8. 132.7 cm^2
9. 162.8 yd^2
10. 551.27 mm^2
11. 1,386 in^2
12. 3.8 cm^2

Practice C

1. 113.0 cm^2
2. 283.4 in^2
3. 38.5 yd^2
4. 6.2 in^2
5. 45.3 cm^2
6. 19.6 ft^2
7. $C \approx 25.1$ in.; $A \approx 50.2$ in^2
8. $C \approx 44.0$ cm; $A \approx 153.9$ cm^2
9. $C \approx 81.6$ ft; $A \approx 530.7$ ft^2
10. $r \approx 8$ m
11. $r \approx 2$ in.
12. $r \approx 10$ yd
13. 803.8 ft^2
14. 1.2 cm^2

Review for Mastery

1. 9; 9; 9; 254.34; 254.3
2. 5; 5; 5; 78.5; 78.5
3. 379.9 yd^2
4. 28.3 m^2
5. 113.0 ft^2
6. 530.7 yd^2

Challenge

1. about 5.4 yd^2
2. about 17.4 ft^2
3. about 28.1 in^2
4. about 340.6 cm^2
5. about 69.7 mm^2
6. about 56.5 in^2
7. about 27.5 m^2
8. about 31.0 ft^2
9. about 50.2 in^2

Problem Solving

1. 577.8 mm^2
2. 447.6 mm^2
3. 15,747.1 in^2
4. 95 in^2
5. D
6. F
7. C
8. G

Reading Strategies

1. a parallelogram
2. $A = bh$
3. half the circumference
4. the radius
5. $A = \frac{1}{2}Cr$

Puzzles, Twisters & Teasers

1. 28.26 m^2
2. 50.24 in^2
3. 78.5 in^2
4. 200.96 cm^2
5. 314 yd^2
6. 32.15 yd^2
7. 63.59 ft^2
8. 176.63 cm^2
9. 3.8 m^2

A L I G H T H O U S E

LESSON 9-6

Practice A

1. C
2. G
3. 17 ft^2
4. 30.28 m^2
5. 174 ft^2
6. 84 m^2
7. 158.13 ft^2
8. 288 m^2
9. 189.25 ft^2

Practice B

1. 21 ft^2
2. 24 ft^2
3. 90 ft^2
4. 208 m^2
5. 140 ft^2
6. 23.13 m^2
7. 100 ft^2
8. 33.28 m^2
9. 57.12 m^2

Practice C

1. 22 ft^2
2. 30 ft^2
3. 104 ft^2
4. 223.4 m^2
5. 60.75 m^2
6. 258.39 m^2
7. $A = 52$ units^2; $P = 36$ units

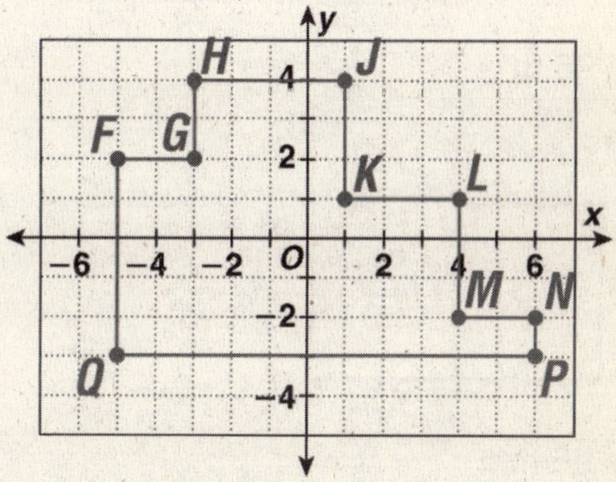

Review for Mastery

1. 9; $1\frac{1}{2}$; $\frac{1}{2}$; 1;

   9; $1\frac{1}{2}$; $\frac{1}{2}$; 1; 12
2. 32; 6; 32; 6; 38

Challenge

1.

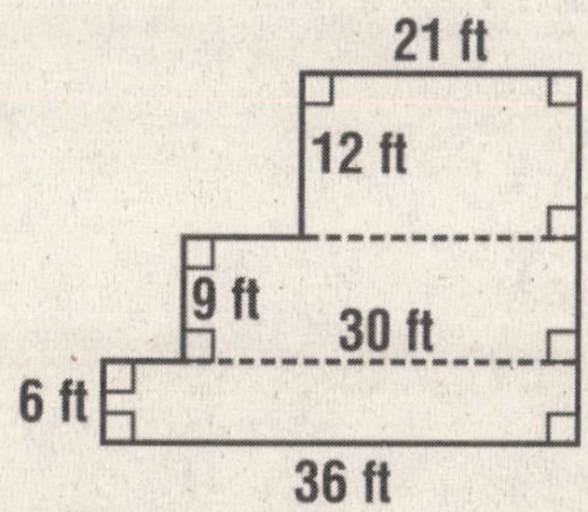

738 ft^2;

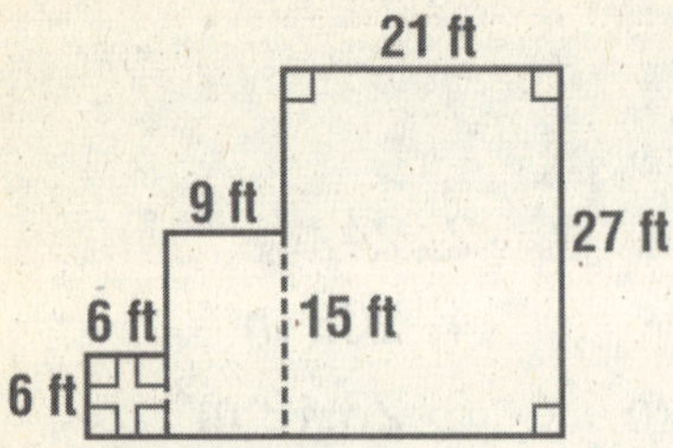

738 ft^2

2.

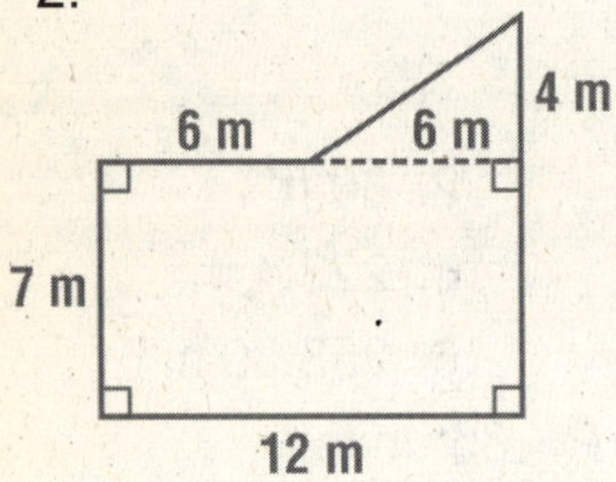

96 m^2;

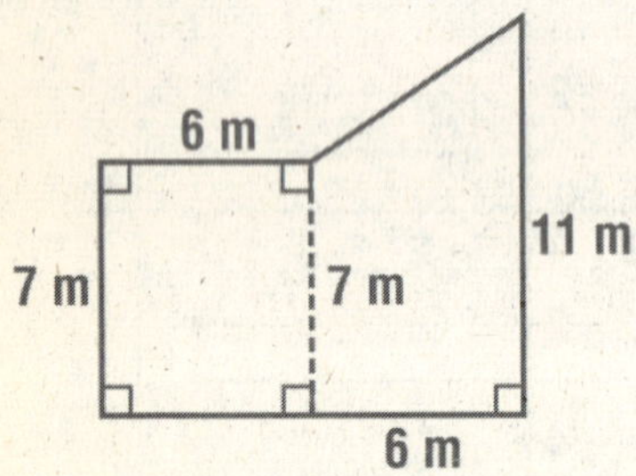

96 m^2

3.

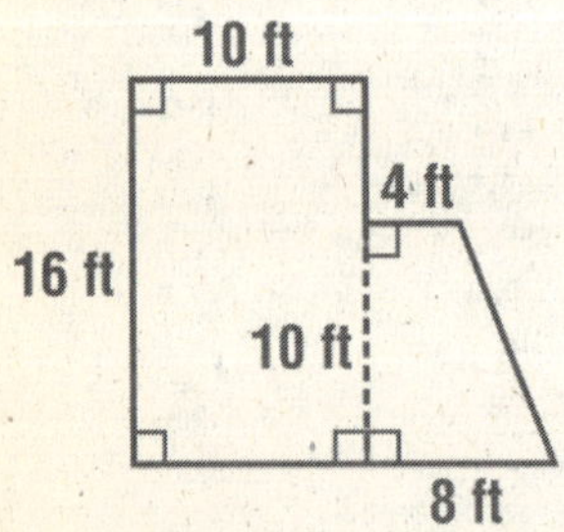

220 ft^2;

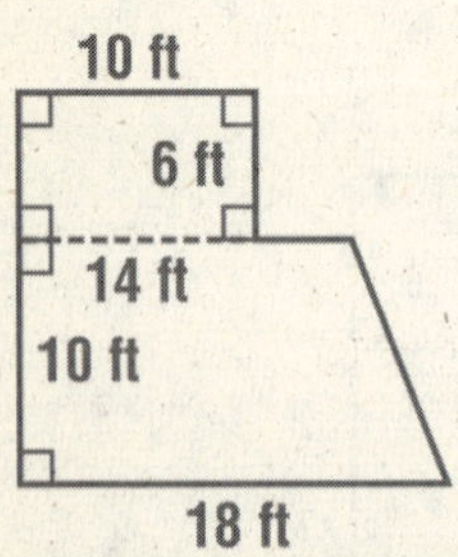

220 ft^2

Problem Solving

1. Possible answer: Divide the figure into a rectangle and a right triangle; 325 ft^2
2. \$1,260
3. 356.48 cm^2
4. 22.5 $in.^2$
5. B
6. H

Reading Strategies

1. a rectangle and a right triangle
2. Use the formula, area = length · width.
3. 80 ft^2
4. Use the formula, area = $\frac{1}{2}$ · base · height.
5. 264 ft^2
6. Add the area of the rectangle to the area of the right triangle
7. 344 ft^2

Puzzles, Twisters & Teasers

1. D
2. N
3. R
4. A
5. O
6. L
7. Y
8. F

A D R A G O N F L Y

LESSON 9-7

Practice A

1. 25
2. 100
3. 121
4. 169
5. 49
6. 144
7. 15
8. 6
9. 9
10. 17
11. 14
12. 22
13. C
14. F
15. B
16. E
17. D
18. A
19. K
20. J
21. G
22. L

23. H
24. I
25. 11 in.

Practice B

1. 36
2. 361
3. 169
4. 16
5. 100
6. 196
7. 400
8. 324
9. 17
10. 7
11. 16
12. 9
13. 11
14. 25
15. 24
16. 30
17. 3
18. 6
19. 10
20. 7
21. 12
22. 4
23. 8
24. 5
25. 6
26. 14
27. 11
28. 9
29. 2
30. 6
31. 9
32. 15
33. 9 ft
34. 12 in.
35. 7 ft
36. 20 cm

Practice C

1. 324
2. 0.49
3. 729
4. 15.21
5. 2,500
6. 0.0961
7. $\frac{1}{25}$
8. $\frac{49}{64}$
9. 28
10. 22
11. 15
12. 31
13. 7
14. 9
15. 3
16. 5
17. 8
18. 4
19. 10
20. 11
21. 12
22. 14
23. 16
24. 13
25. 21
26. 19
27. 24
28. 32
29. 9
30. 10
31. 1
32. 52
33. 23
34. 173
35. 144 in^2
36. 60 yd
37. 55 in.

Review for Mastery

1. 5
   5; 5
   25
2. 9
   9; 9
   81
3. 169
4. 256
5. 121
6. 49
7. 49
   7
   7
8. 81
   9
   9

Challenge

1.

Powers of 3	
3^1	3
3^2	9
3^3	27
3^4	81
3^5	243
3^6	729
3^7	2,187
3^8	6,561

Powers of 7	
7^1	7
7^2	49
7^3	343
7^4	2,401
7^5	16,807
7^6	117,649
7^7	823,543
7^8	5,764,801

Powers of 9	
9^1	9
9^2	81
9^3	729
9^4	6,561
9^5	59,049
9^6	531,441
9^7	4,782,969
9^8	43,046,721

2. 3, 9, 7, 1 repeating
3. 3; 7; 9
4. 7, 9, 3, 1 repeating
5. 9; 1; 7
6. Possible answer: No; The powers of 9 have only two repeating digits.
7. 1; 1; 1; 1
8. Possible answer:1
9. Except for 5^1, the last 2 digits always form 25.
10. They are all 6's.

Problem Solving

1. 19.6 m
2. 122.5 m
3. 9 pies
4. 128 ft
5. C
6. J
7. B
8. F

Reading Strategies

1. It has two factors that are the same.
2. five squared
3. 25
4. the square root of 25

Puzzles, Twisters & Teasers

P	E	R	F	E	C	T	R	C	V	B
O	V	U	J	E	R	V	A	L	U	E
W	G	T	Y	U	I	L	D	S	Q	W
E	X	P	R	E	S	S	I	Y	E	R
R	M	J	I	O	Q	I	C	M	Z	X
O	V	B	N	V	U	G	A	B	O	I
O	I	O	P	L	A	N	L	O	P	U
T	B	H	V	I	R	U	S	L	L	I
S	A	M	O	D	E	L	A	S	E	C

virus

LESSON 9-8

Practice A

1. B
2. G
3. B
4. H
5. C
6. G
7. about 13.7 ft

Practice B

1. 15 in.
2. 8 cm
3. 26 ft
4. 24 m
5. 5 yd
6. 10 m
7. 16 mm
8. 30 in.
9. 15 yd
10. 16 ft
11. 14.1 in.
12. 14.4 mi
13. 32 m

Practice C

1. 17 in.
2. 25 cm
3. 24 ft
4. 7 cm
5. 35 yd
6. 30 m
7. 9.8 ft
8. 8.9 m
9. 13.2 ft
10. 12.2 m
11. no
12. yes
13. no
14. yes
15. 18.3 ft
16. 11.3 cm

Review for Mastery

1. 6; 8
   36; 64
   100
   10 ft
2. 5; 13
   25; 169
   144
   12 cm
3. 15 m
4. 16 in.
5. 20 mm
6. 18 yd

Challenge

1.

Multiply by 3.	(9, 12, 15)
Multiply by 5.	(15, 20, 25)
Multiply by 10.	(30, 40, 50)
Divide by 5.	$\left(\frac{3}{5}, \frac{4}{5}, 1\right)$

2.

Multiply by 2.	(10, 24, 26)
Multiply by 3.	(15, 36, 39)
Multiply by 8.	(40, 96, 104)
Divide by 10.	$\left(\frac{1}{2}, \frac{6}{5}, \frac{13}{10}\right)$

3.

Divide by 4.	$\left(\frac{7}{4}, 6, \frac{25}{4}\right)$
Multiply by 2.	(14, 48, 50)
Multiply by 4.	(28, 96, 100)
Multiply by 12.	(84, 288, 300)

4.

Divide by 2.	$\left(4, \frac{15}{2}, \frac{17}{2}\right)$
Multiply by 3.	(24, 45, 51)
Multiply by 4.	(32, 60, 68)
Multiply by 20.	(160, 300, 340)

5. Possible answer: $15^2 + 20^2 = 25^2$; $225 + 400 = 625$
6. Possible answer: $10^2 + 24^2 = 26^2$; $100 + 576 = 676$
7. Possible answer: $28^2 + 96^2 = 100^2$; $784 + 9{,}216 = 10{,}000$

Problem Solving

1. 51 ft
2. 32 ft
3. 150 cm^2
4. less, because $\sqrt{13{,}600}$ is less than 200
5. C
6. J
7. B
8. F

Reading Strategies

1. 90
2. legs
3. hypotenuse
4. 9 cm
5. 16 cm
6. 25 cm
7. 25 cm
8. They are the same.

Puzzles, Twisters & Teasers

1. 13 cm
2. 15 m
3. 20 m
4. 17 ft
5. 5 cm
6. 10 in.
7. 5.8 m
8. 4.2 yd
9. 127.3 cm

BIGGER TRUNK